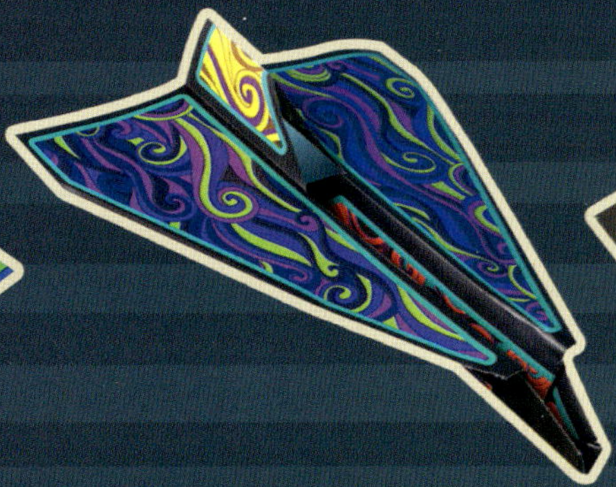

PAPER AIRPLANE KIT

FOLDING INSTRUCTIONS

Publications International, Ltd.

Written by Brandon Myers

Cover and interior art by Nick LaShure, Shutterstock.com

Louis Weber, CEO
Publications International, Ltd.
8140 Lehigh Avenue
Morton Grove, IL 60053

ISBN: 978-1-63938-675-8

Manufactured in China.

8 7 6 5 4 3 2 1

CONTENTS

Classroom *Classic*

This easy design is the first one we usually learn. The simple, stable design makes it easy to throw.

EASY | MEDIUM | HARD

Lay the paper flat with the name facing **down**.

1 Fold paper in half on center line, and open up the paper so it lies flat.

2 Fold down both corners.

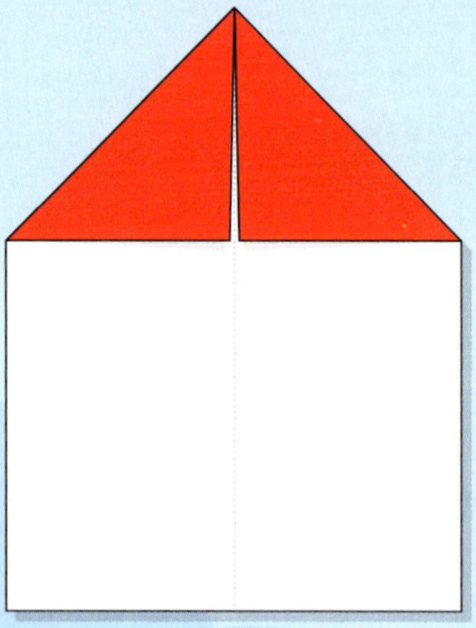

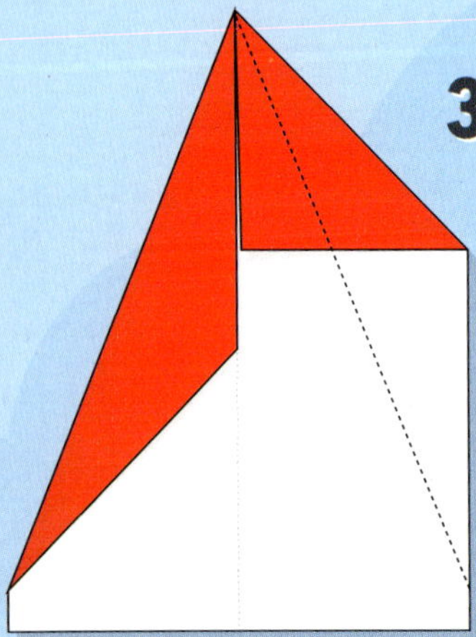

3 Fold down the sides.

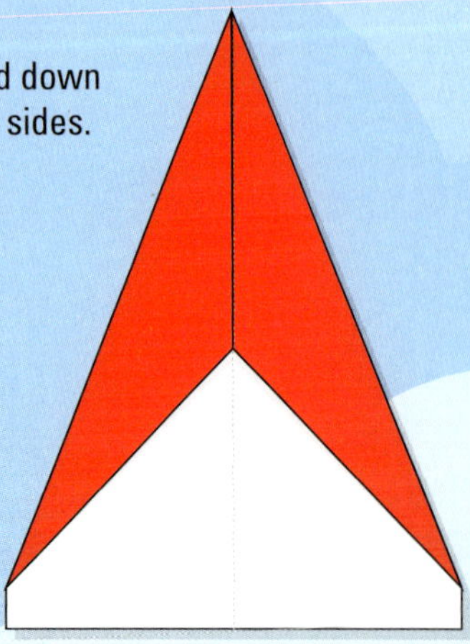

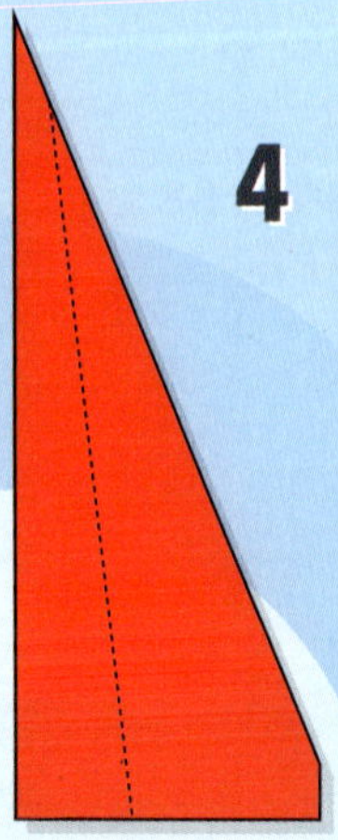

4 Fold the plane in half along the center line. Then fold down each wing.

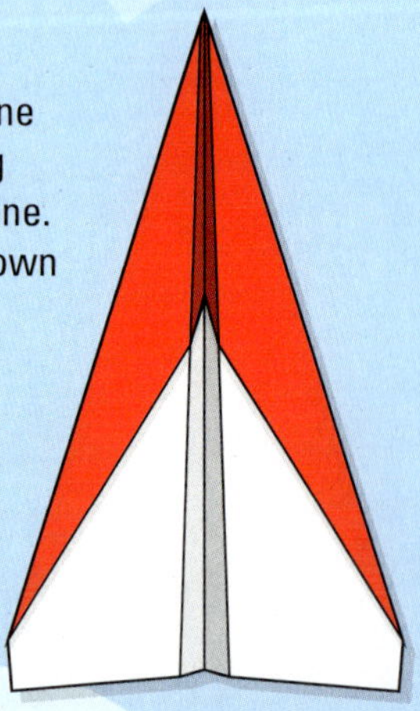

THE NEEDLE

EASY | MEDIUM | HARD

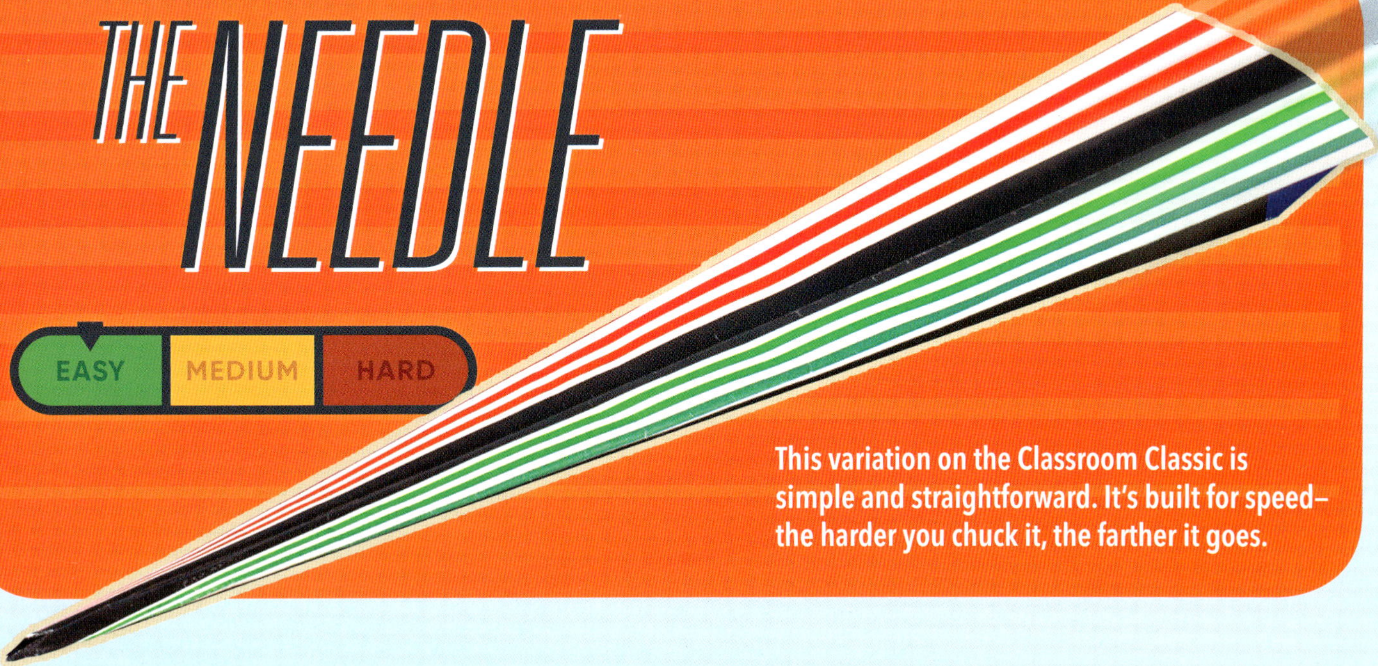

This variation on the Classroom Classic is simple and straightforward. It's built for speed—the harder you chuck it, the farther it goes.

Lay the paper flat with the name facing **down**.

1 Fold paper in half on center line, and open up the paper so it lies flat.

2 Fold down both corners.

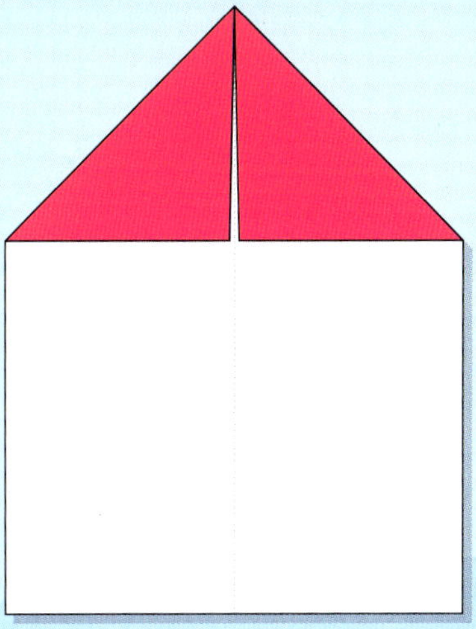

3 Fold down the sides.

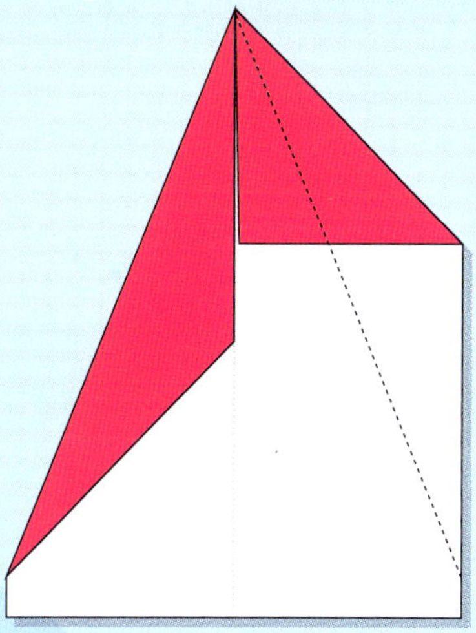

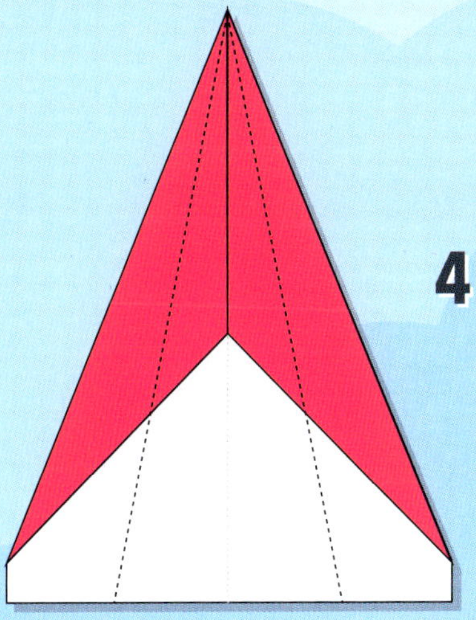

4 Fold down once more!

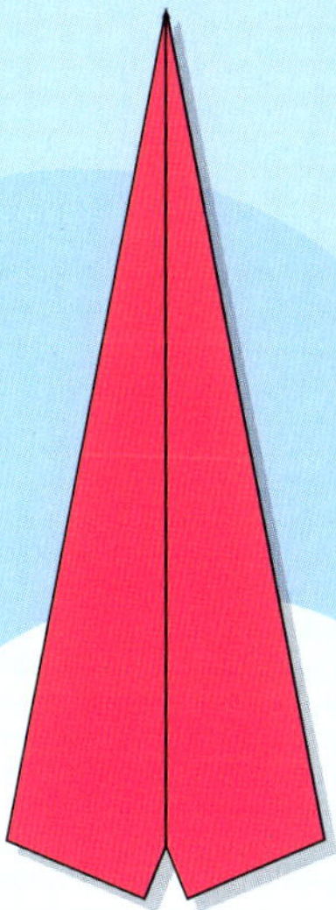

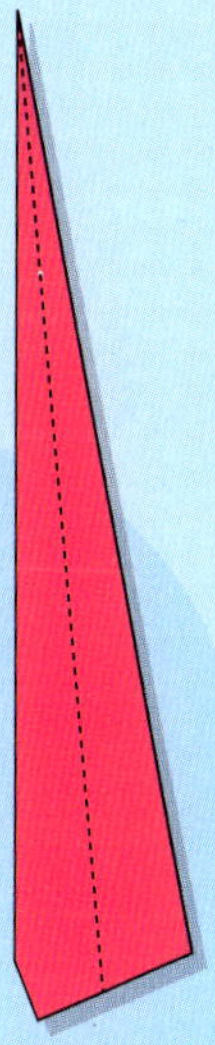

5 Fold the plane in half along the center line. Then fold down each wing.

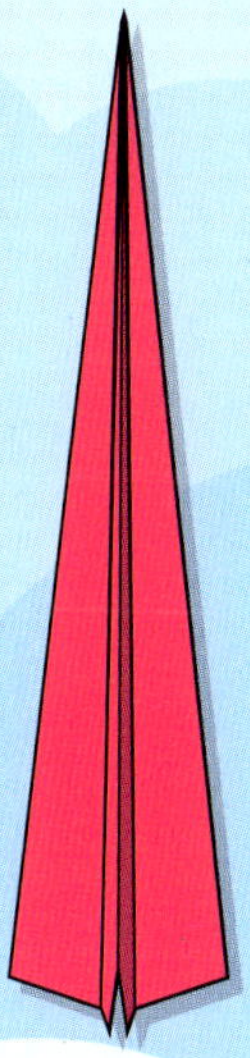

CLOUD DRAGON

This unusual design has a stabilizing forewing known as a *canard*. It may loop and turn a bit, but this glider is capable of drifting long distances before returning gracefully to the earth. Experiment with tape, paper clips, and throwing speed to discover what it can do.

EASY | MEDIUM | HARD

Lay the paper flat with the name facing **up**.

1 Fold paper in half on center line, and open up the paper so it lies flat.

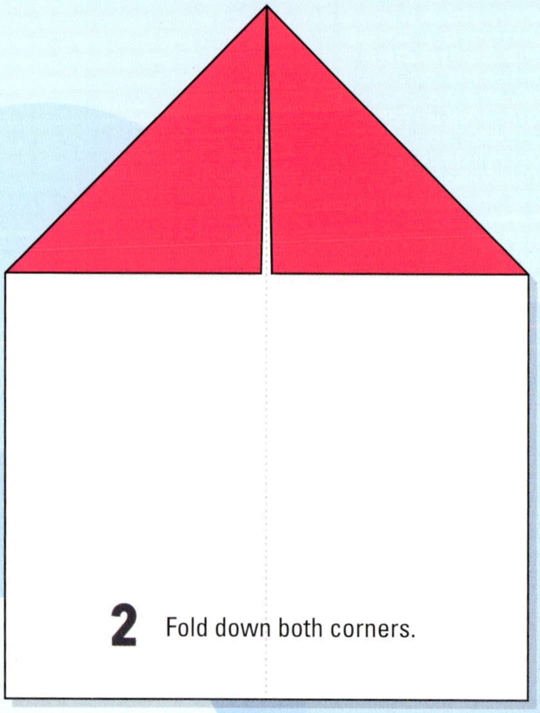

2 Fold down both corners.

3 Turn plane over. Fold down both sides.

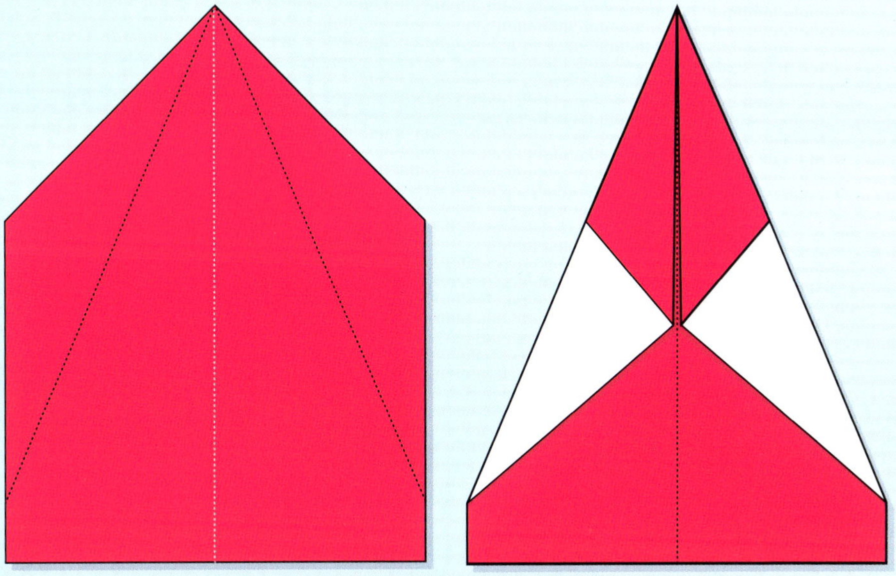

4 Pull out flaps.

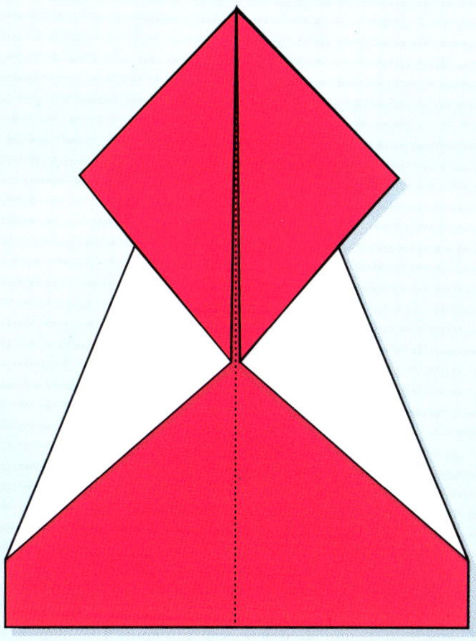

5 Turn plane over.
Fold down top.

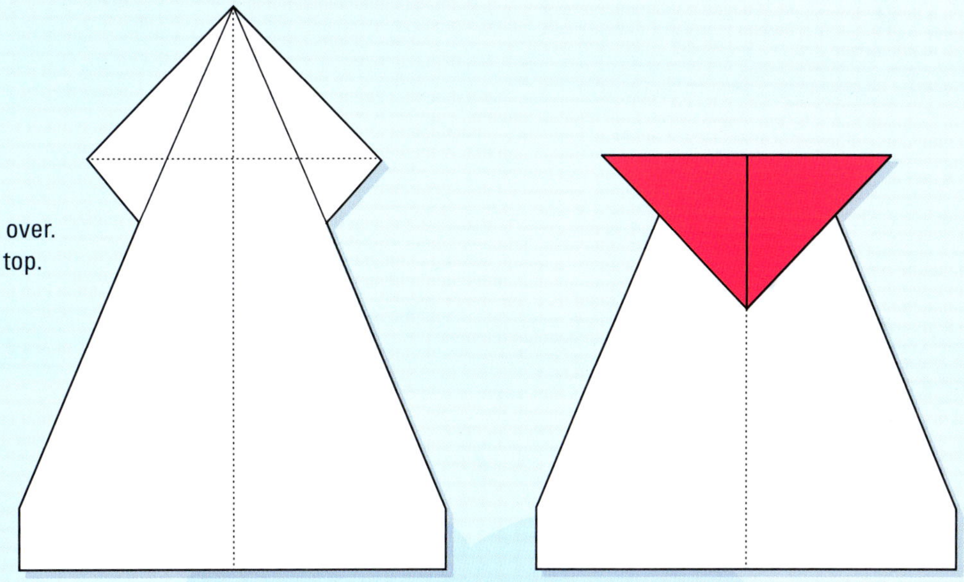

6 Fold plane in half along the center line. Then fold down each wing.

the fly

EASY MEDIUM HARD

It looks like a fly. Does it fly like a fly? You decide. You can create (or fix) a spiralling flight by bending one wing's trailing edge up or down.

Lay the paper flat with the name facing **down**.

1 Fold paper in half on center line, and open up the paper so it lies flat.

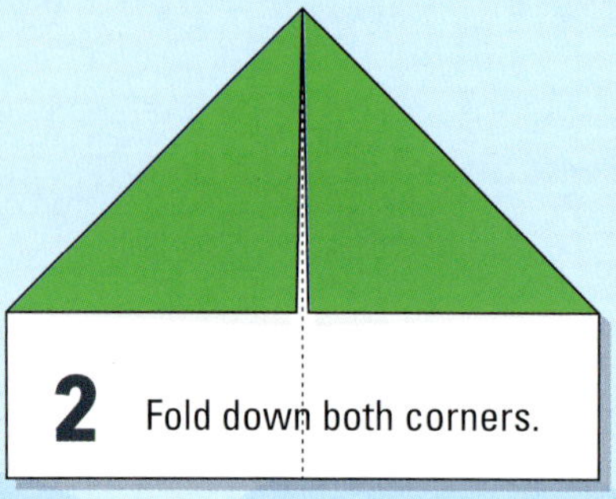

2 Fold down both corners.

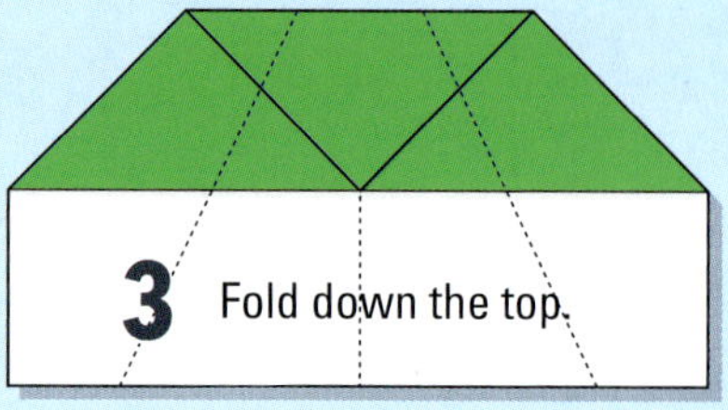

3 Fold down the top.

4 Fold down both sides in to the center line.

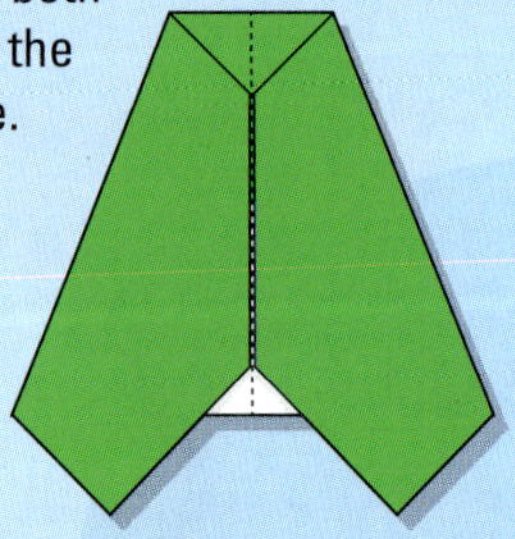

5 Turn the plane over. Fold down the top.

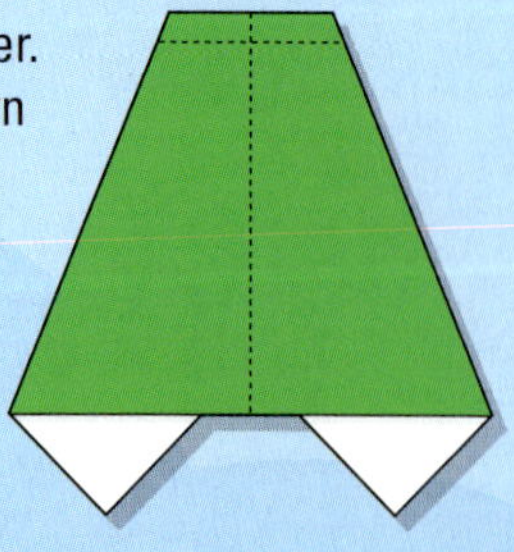

6 Fold the plane in half. Make sure you fold it so that the fold in step 5 is on the outside.

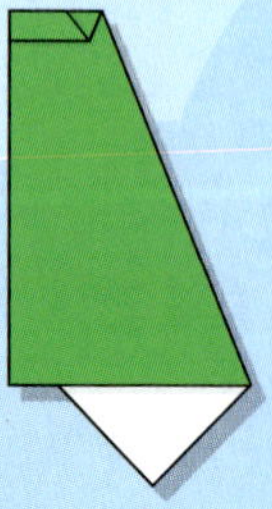

7 Fold down each wing.

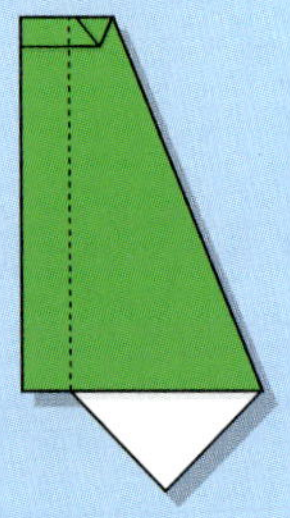

8 Fold up (or down) the winglets.

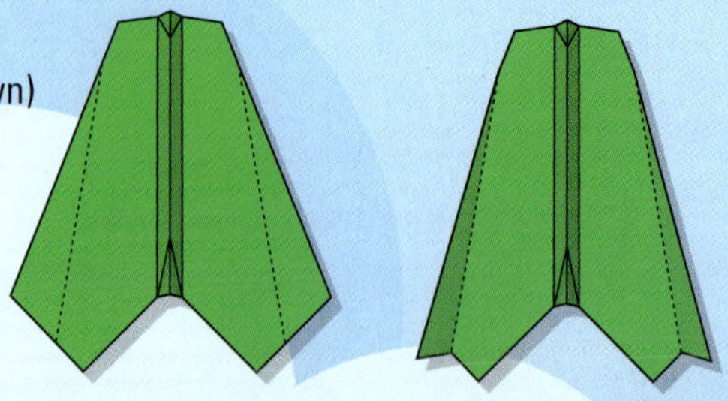

PHOENIX

The Phoenix is an unpredictable bird. It may loop upward or take a fiery dive. Throw it downward and watch it suddenly rise up. You can straighten out its flight by adding one or two paper clips to its nose.

Lay the paper flat with the name facing **down**.

1 Fold paper in half on center line, and open up the paper so it lies flat.

2 Fold down both corners.

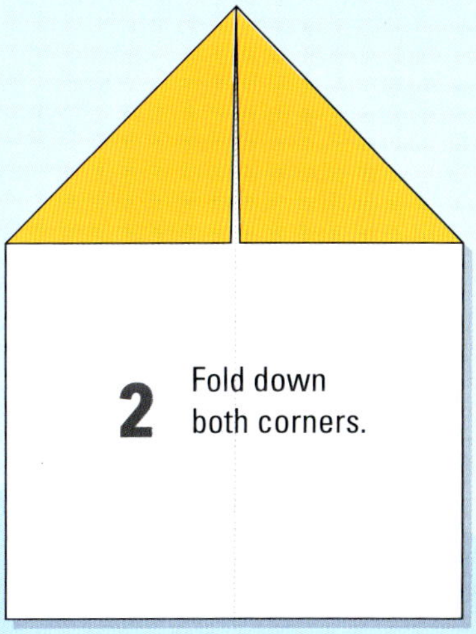

3 Fold down the top, then fold the top corners to meet at the arrows.

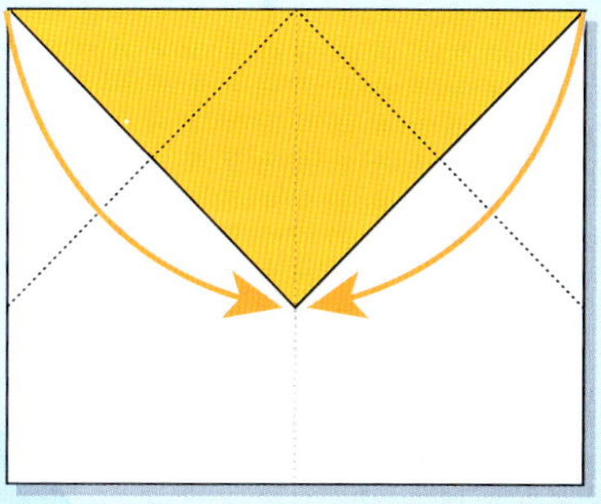

4 Fold plane in half along center line.

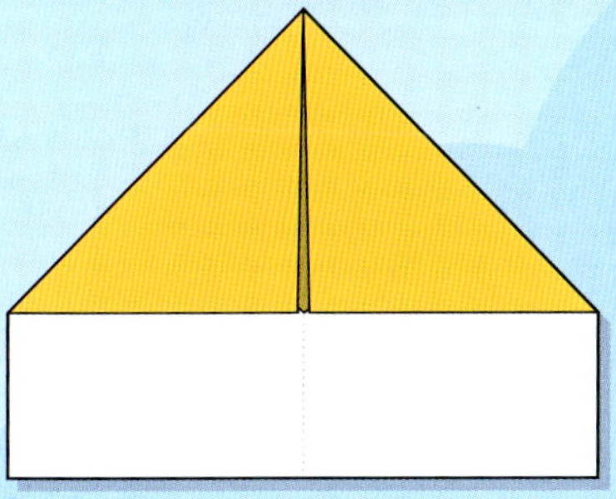

5 Fold down each wing. Cut the tail along the solid cutting line.

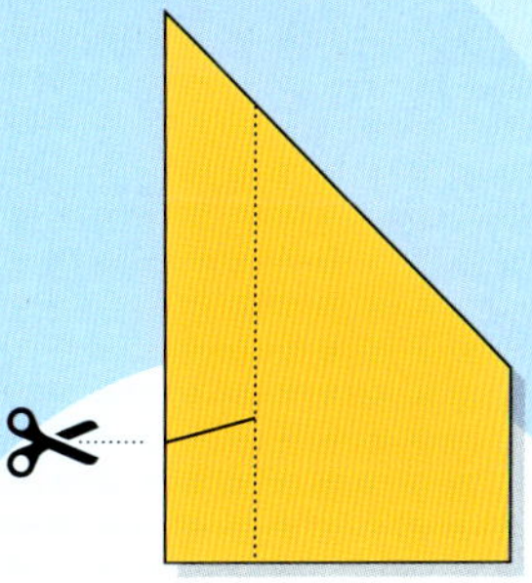

6 Fold up both winglets.

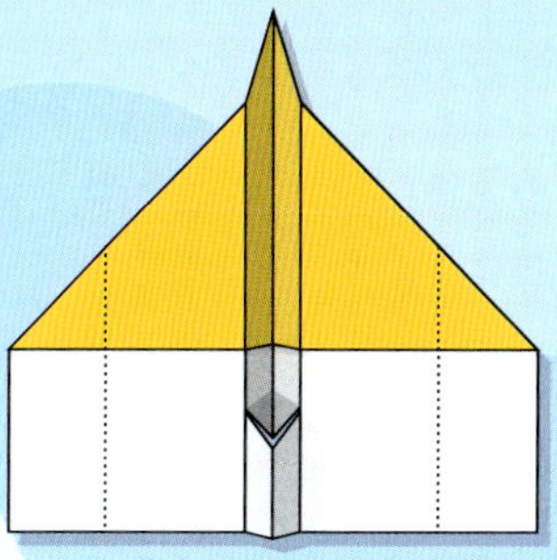

7 Pull the tail up.

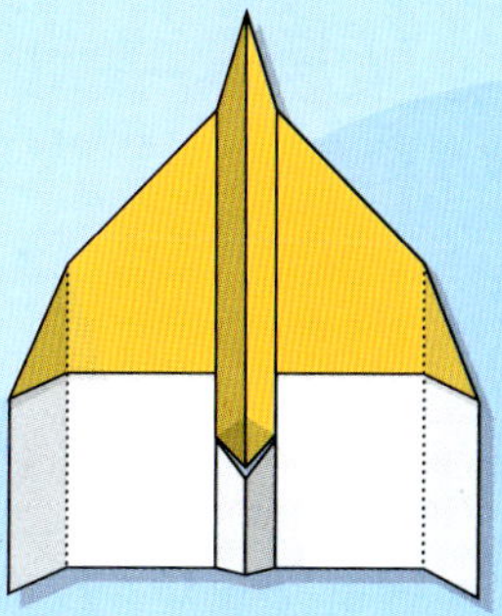

KATANA

This streamlined model slices and spirals through the air. A piece of tape on top of the fuselage will change its flight behavior. If it tends to fly downward, stabilize its pitch by adding a paper clip to the nose.

Lay the paper flat with the name facing **down**.

1 Fold paper in half on center line, and open up the paper so it lies flat.

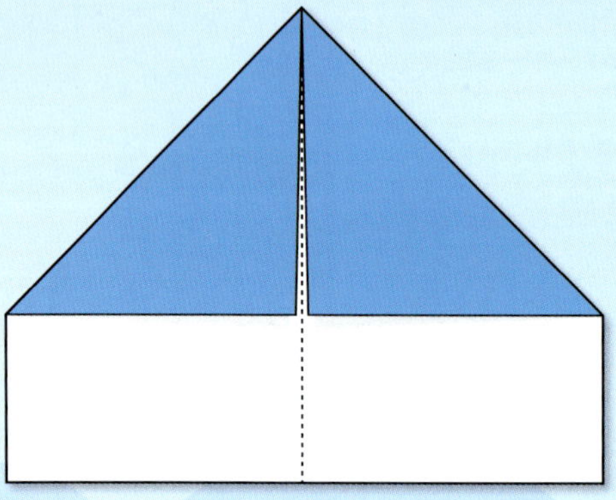

2 Fold down both corners.

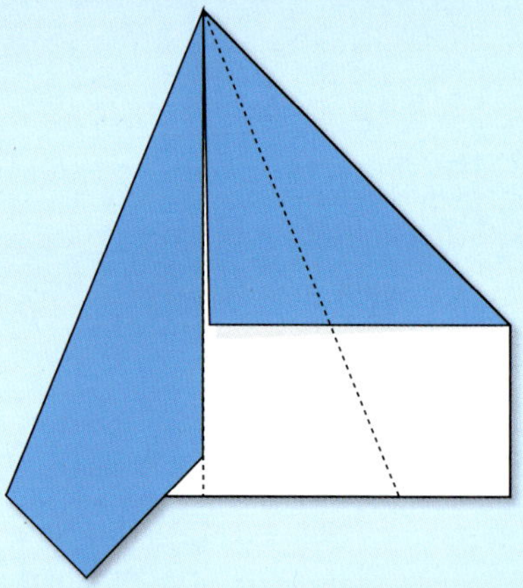

3 Fold down the corners again.

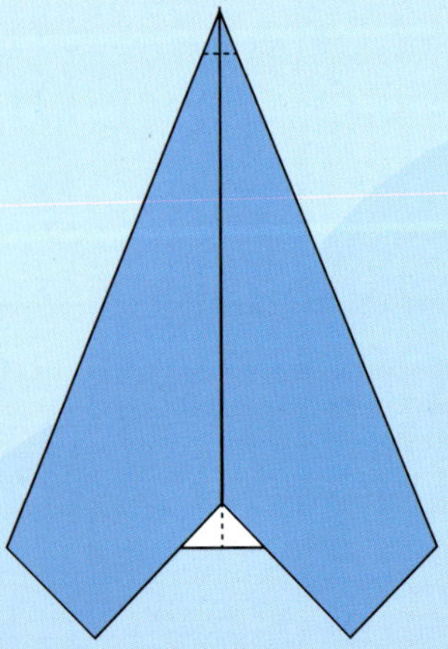

4 Fold down the top point.

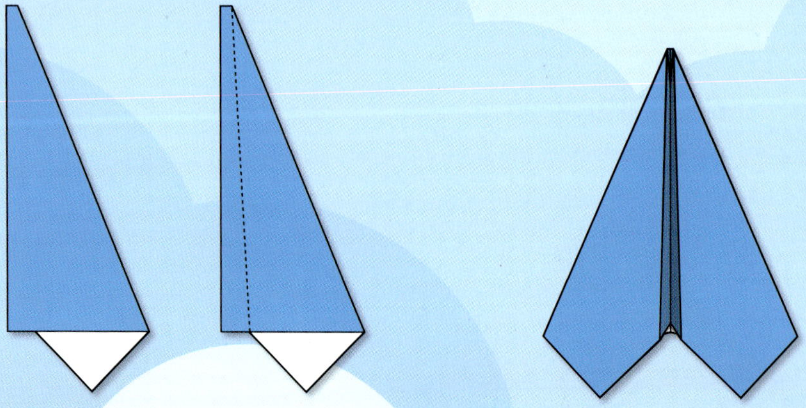

5 Fold the plane in half along the center line. Make sure you fold it so that the fold in step 4 is on the inside. Then fold down each wing.

CLOSE ENCOUNTER

EASY | MEDIUM | HARD

Don't be surprised if this interstellar traveler tries to swoop straight up into space. The saucer-like wings give this craft flights of high strangeness!

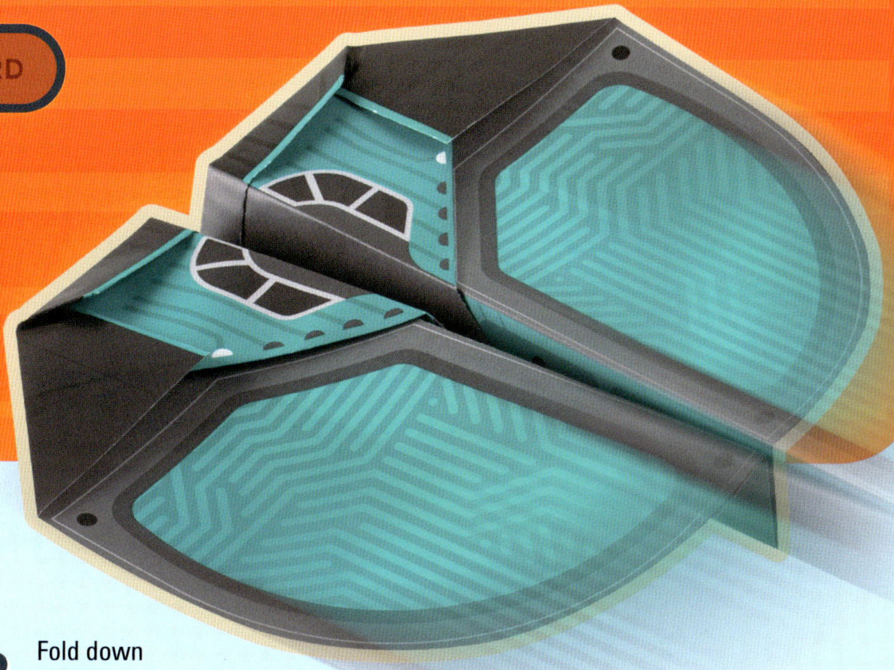

Lay the paper flat with the name facing **down**.

1 Fold paper in half on center line and cut off the edge marked by the solid cutting line. Open up the paper so it lies flat.

2 Fold down both corners.

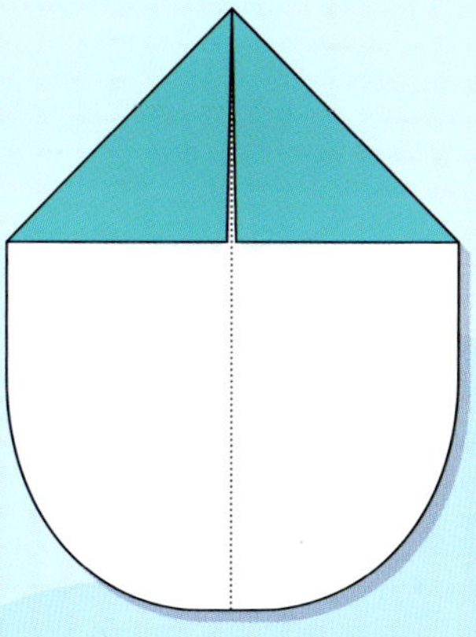

3 Fold down the top.

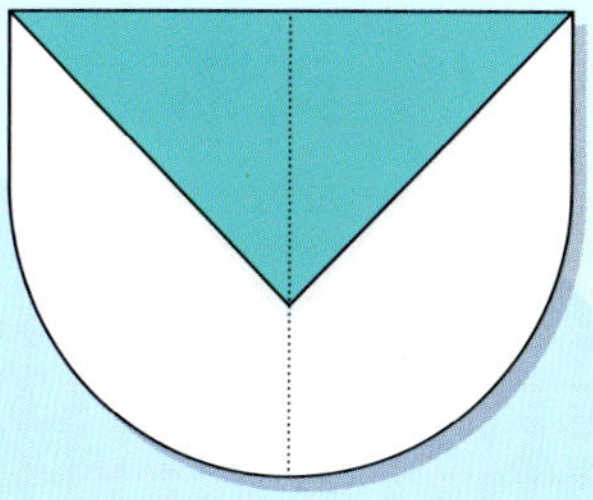

4 Fold down the two corners. The points should fall on the edges of the previous fold.

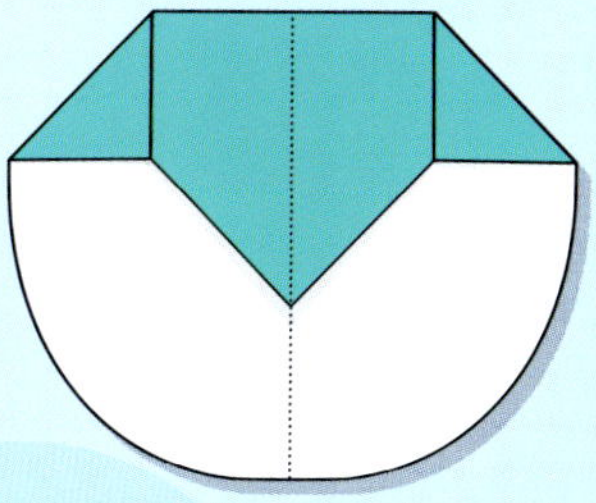

5 Fold the top edge down.

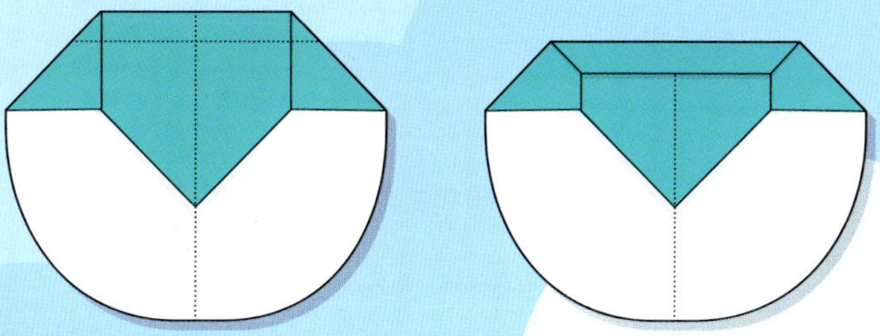

6 Fold the plane in half along the center line. Then fold down each wing.

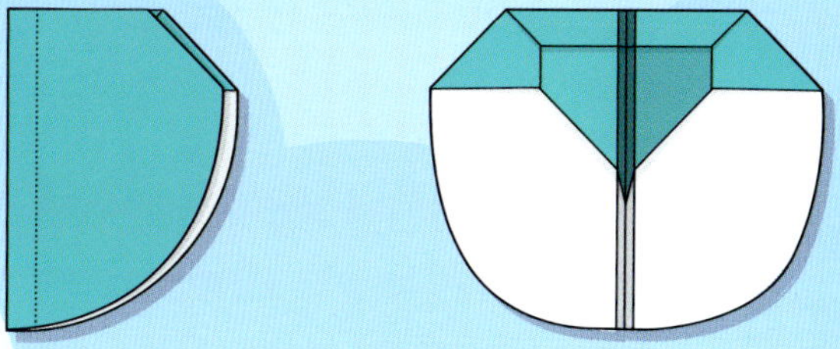

TANGENT

EASY | MEDIUM | HARD

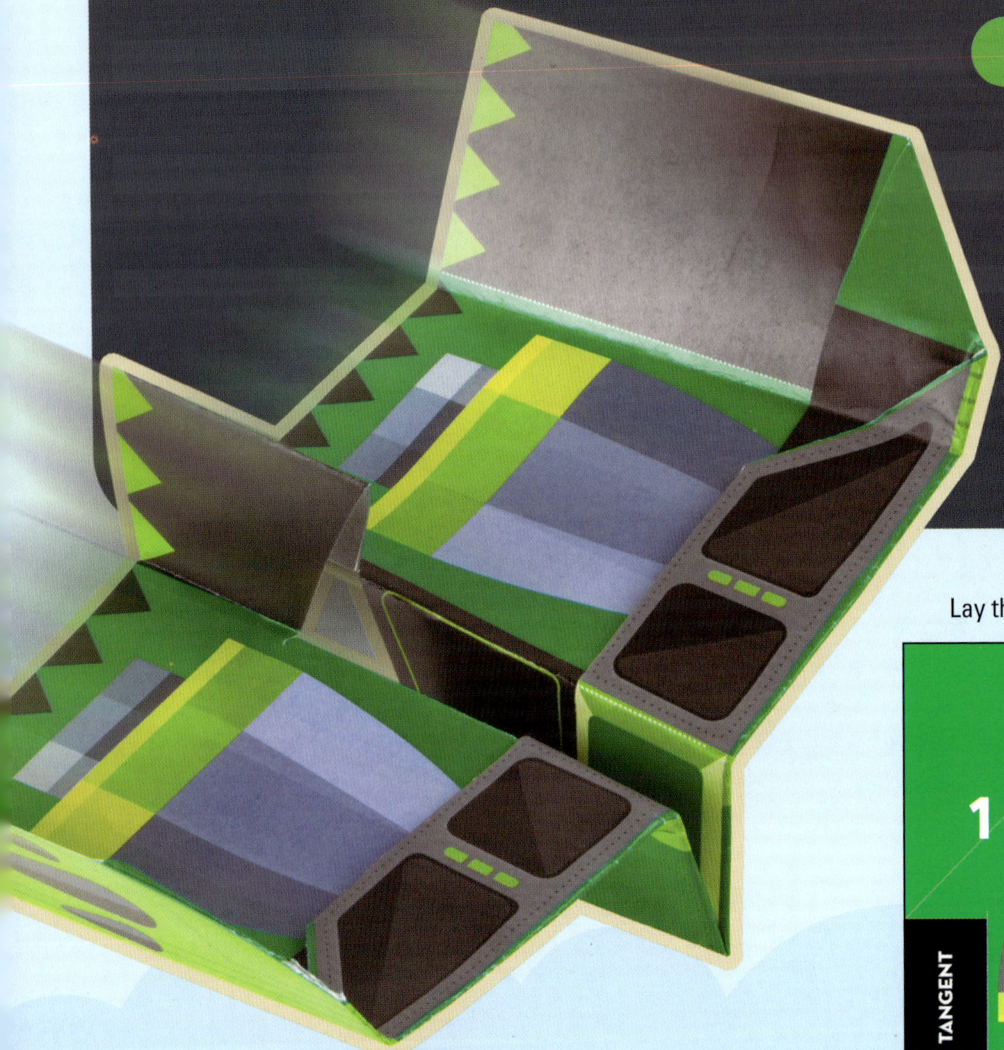

Expect abrupt changes of course with the unpredictable Tangent. It swoops and curves best with a soft throw. You don't need to use tape on this model–it performs better with an open fuselage.

Lay the paper flat with the name facing **up**.

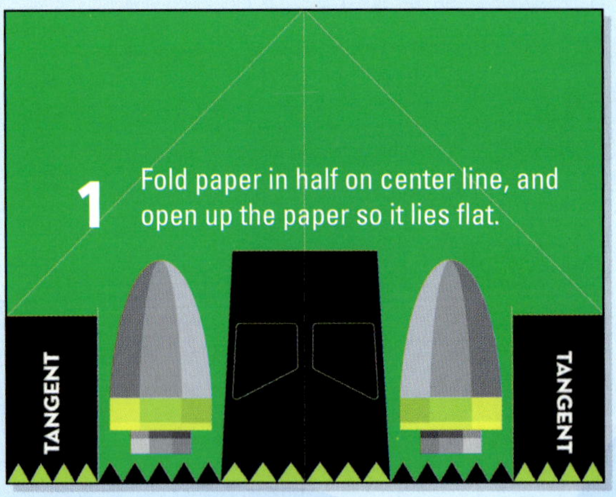

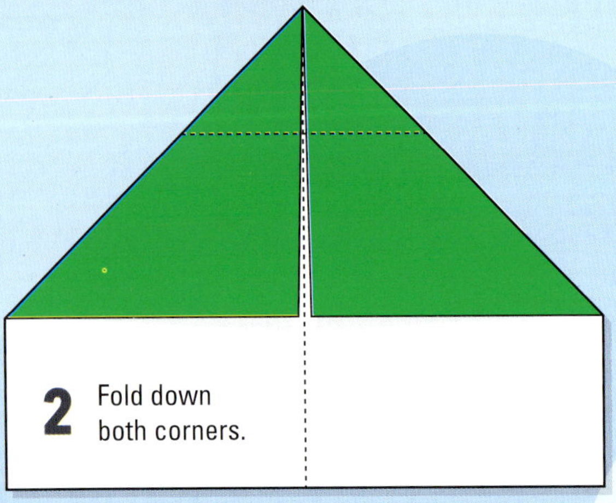

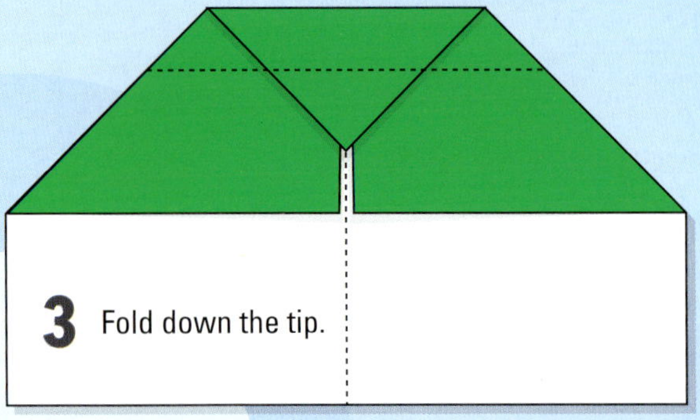

4 Fold the top side down.

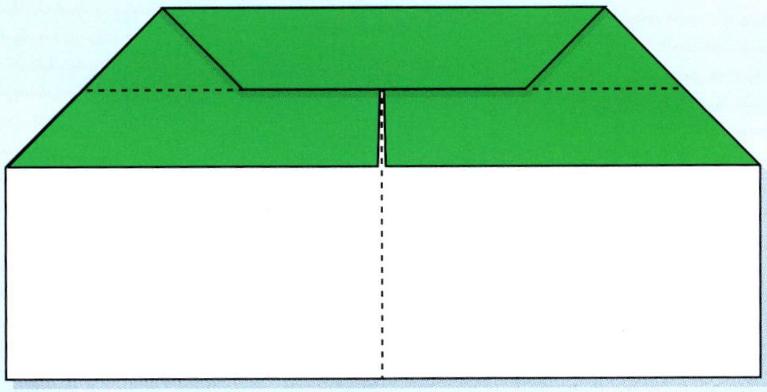

5 Fold the top side down once more.

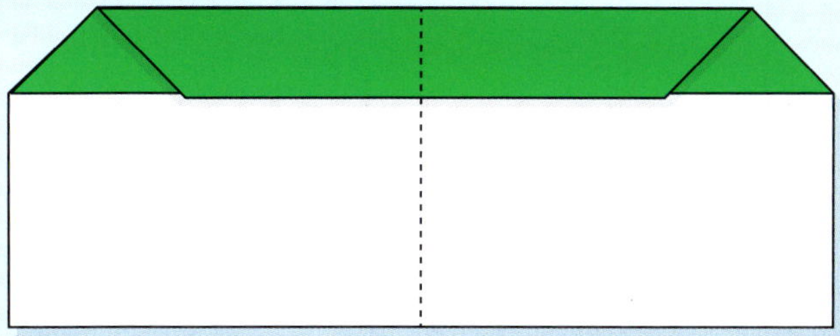

6 Fold the plane in half, with the last several folds on the inside.

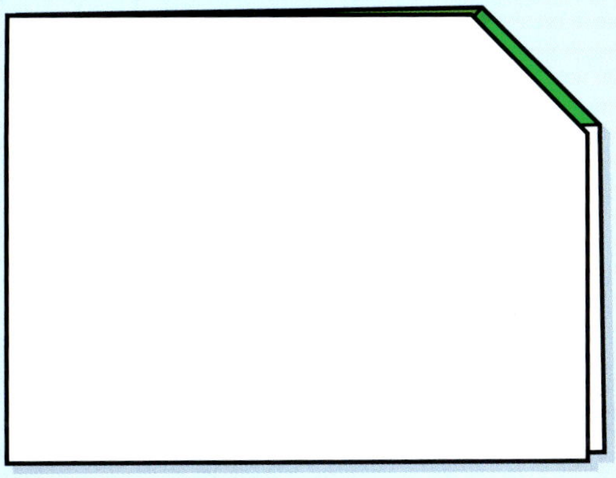

7 Fold down each wing. Cut the tail along the solid cutting line.

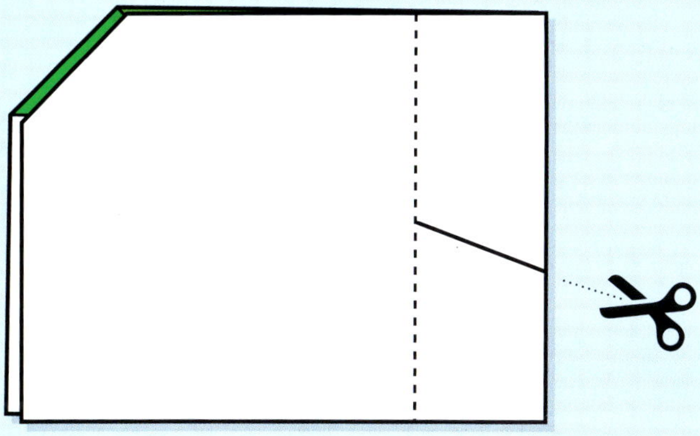

8 Fold up both winglets. Pull the tail up.

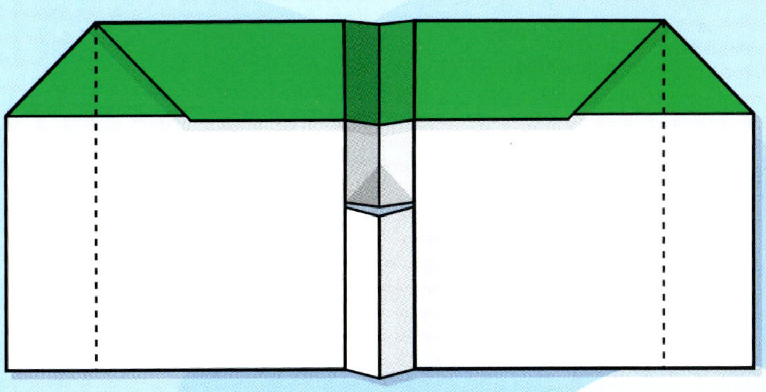

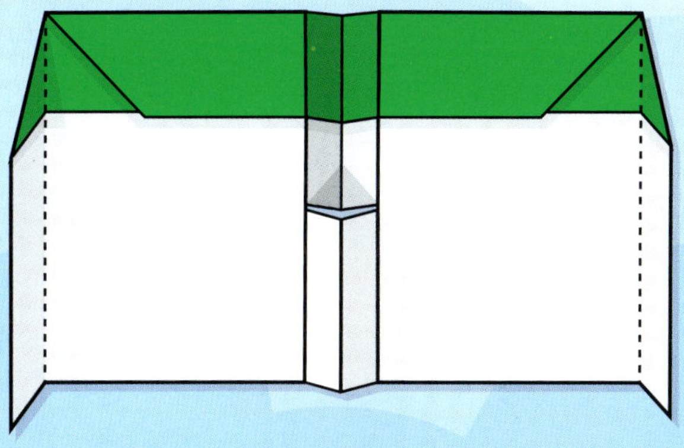

Orbiter

EASY MEDIUM HARD

Keep calm and carry cargo to near-Earth orbit.

Lay the paper flat with the name facing **up**.

1 Fold the paper in half.

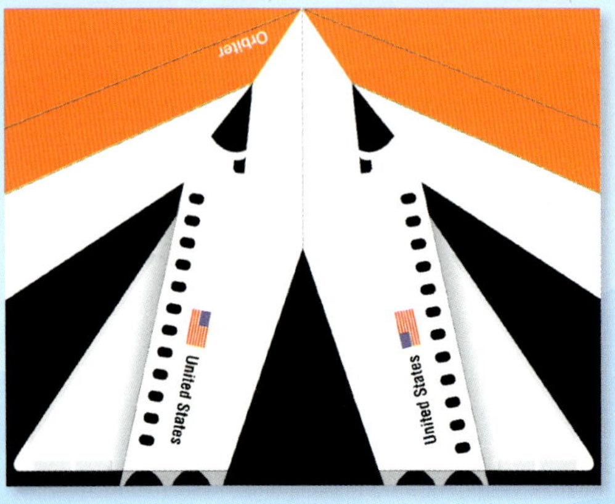

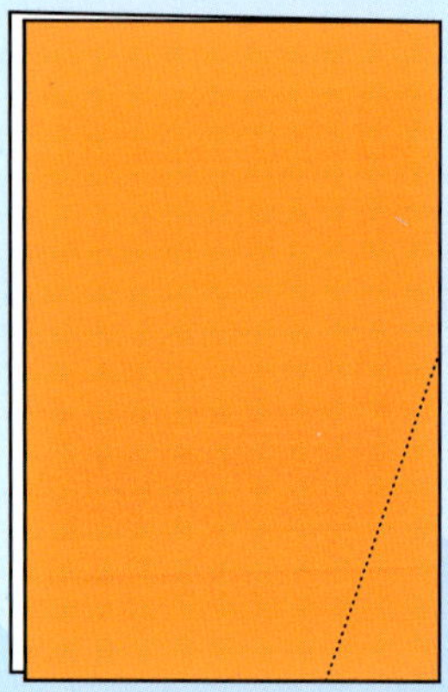

2 Make the crease shown on both sides.

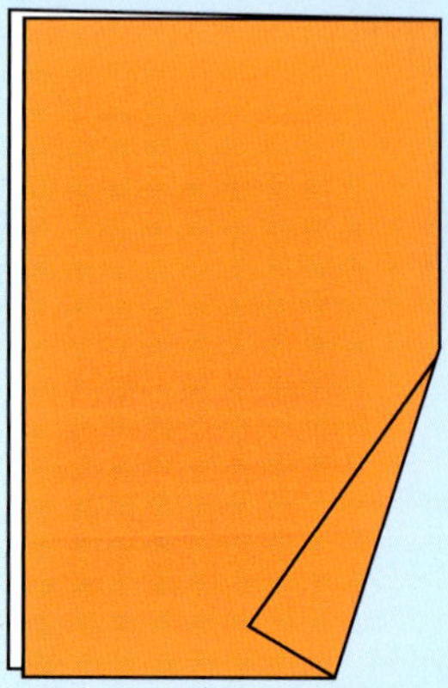

3 Tuck the creased portion inside the folded paper.

4 Fold down the top corner. Turn the plane over and repeat step.

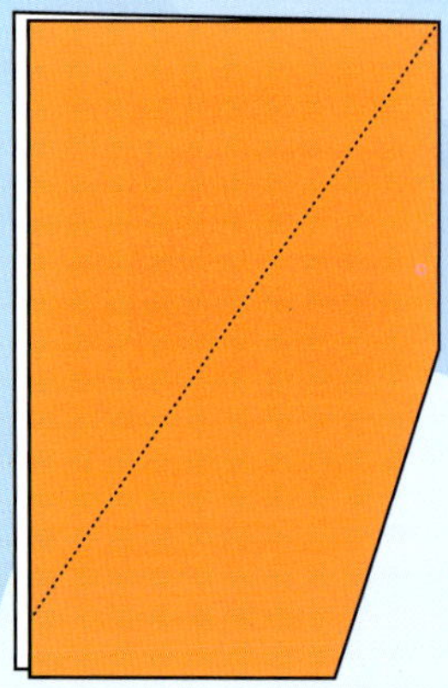

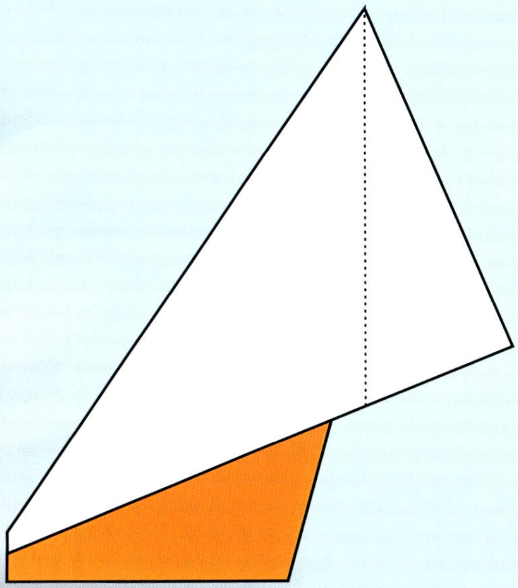

5 Crease the outer corner. Use the crease to tuck the flap inward. Repeat step for the other side.

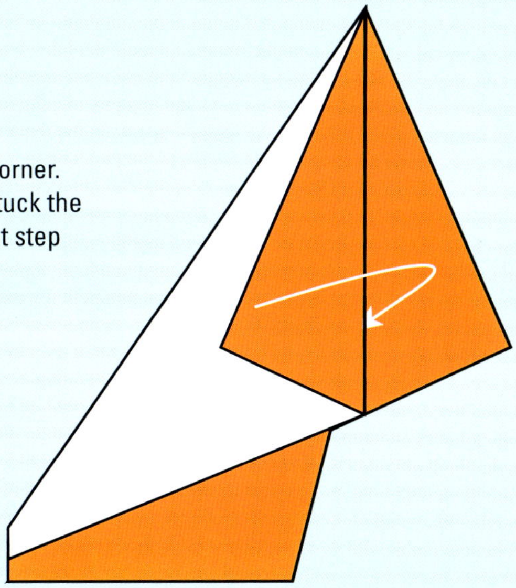

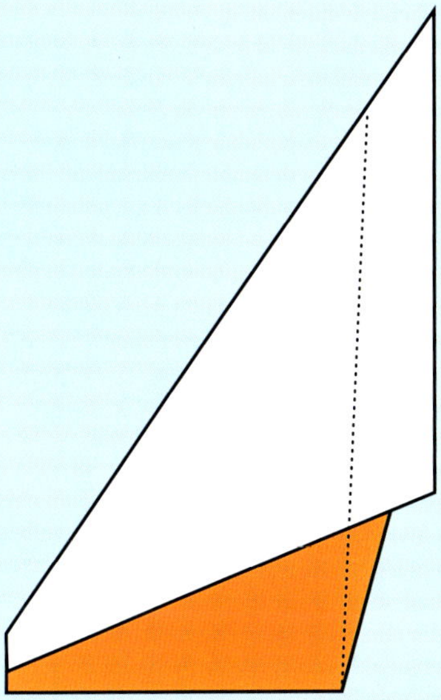

6 Fold down each wing.

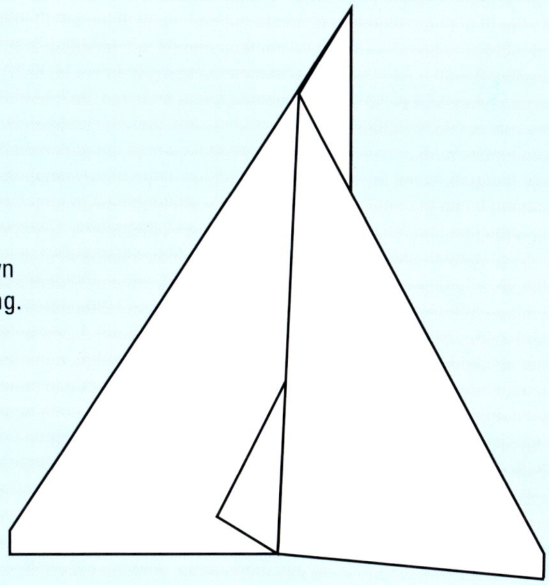

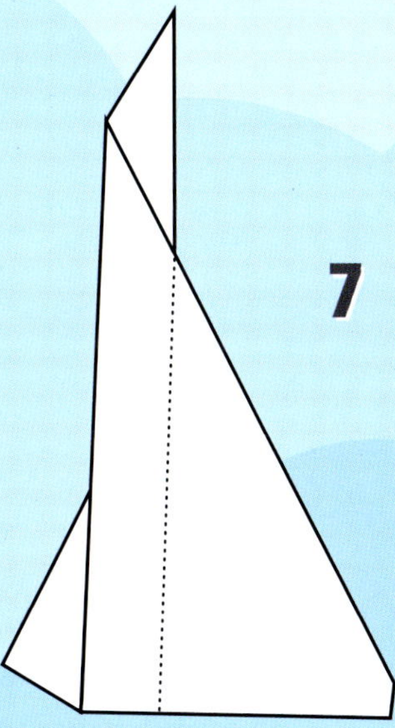

7 Fold the wings again, but upward this time.

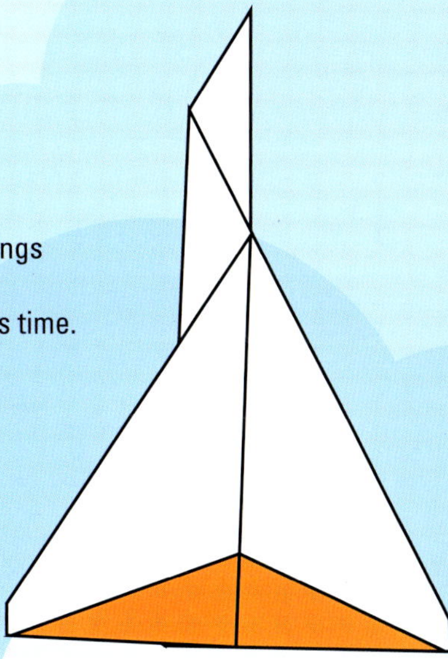

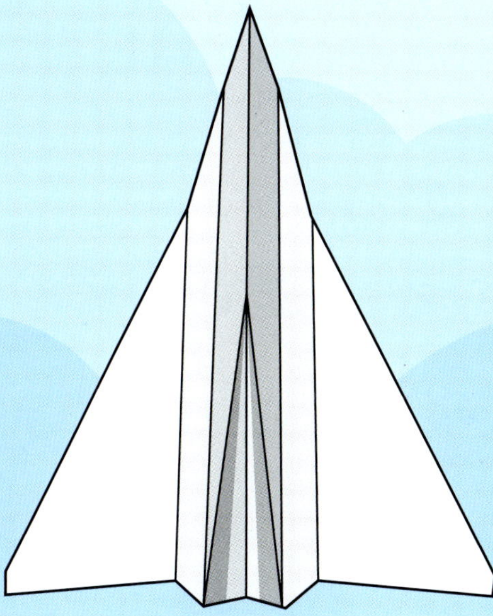

STEALTH GLIDER

This glider likes to swoop through the air at medium speed. You can modify its flight behavior by taping the fuselage or top flaps.

EASY | **MEDIUM** | HARD

Lay the paper flat with the name facing **down**.

1 Fold paper in half on center line and open up the paper so it lies flat.

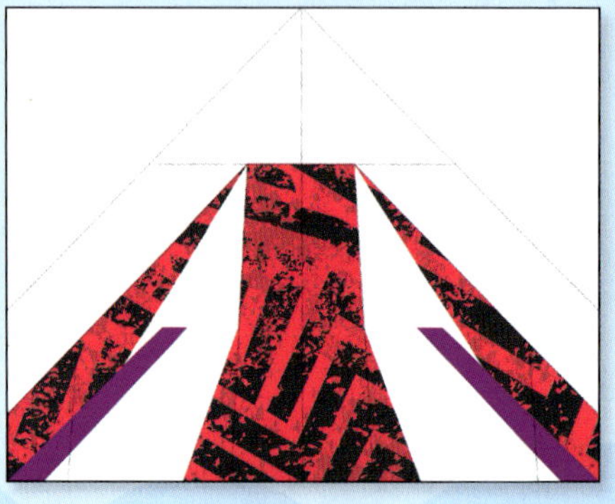

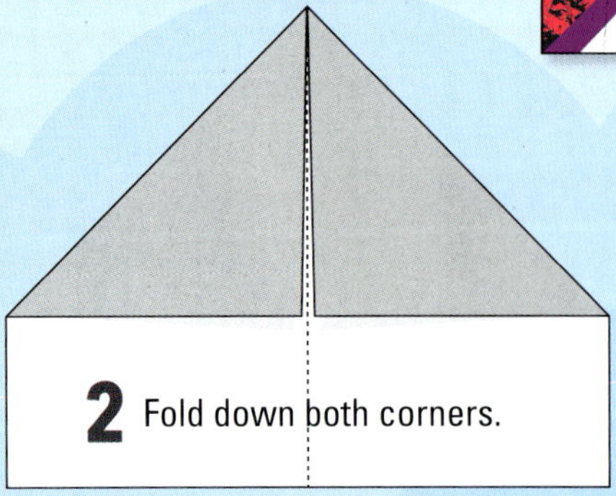

2 Fold down both corners.

3 Fold down the tip.

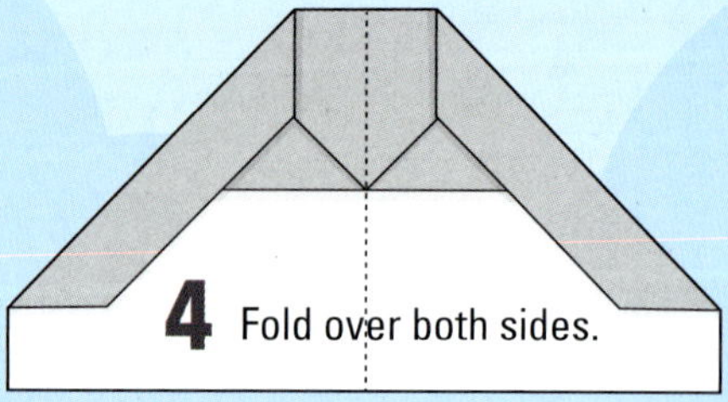

4 Fold over both sides.

5 Fold the plane in half along the center line. Then fold down each wing.

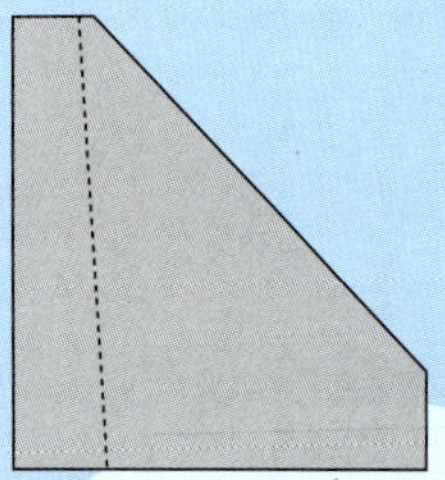

6 Fold up both winglets.

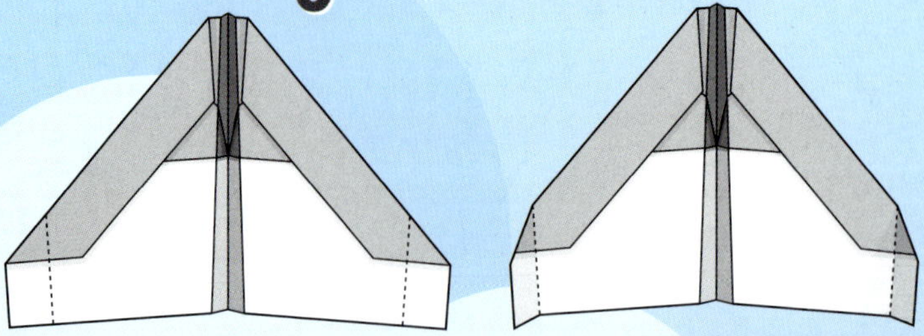

REGAL EAGLE

This soaring bird's streamlined design can handle flights at any speed. After folding the plane, leave the tail and fuselage open for a long, smooth flight.

EASY | **MEDIUM** | HARD

Lay the paper flat with the name facing **down**.

1 Fold paper in half on center line and open up the paper so it lies flat.

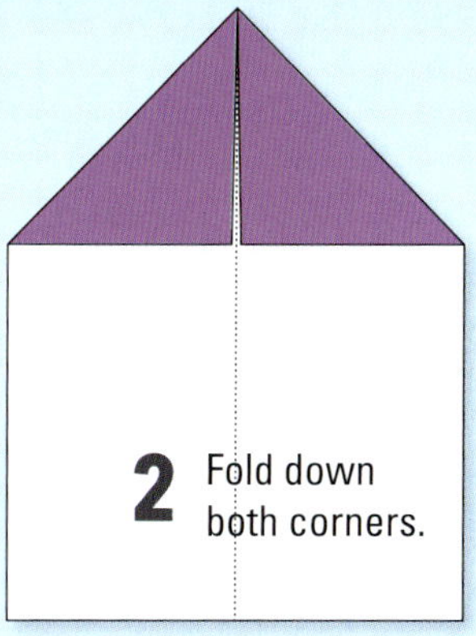

2 Fold down both corners.

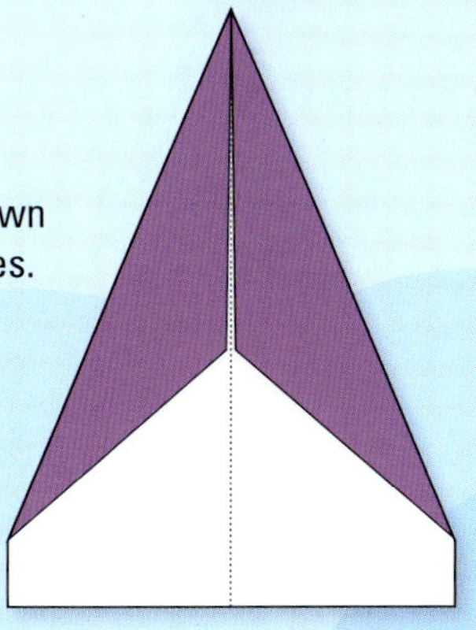

3 Fold down the sides.

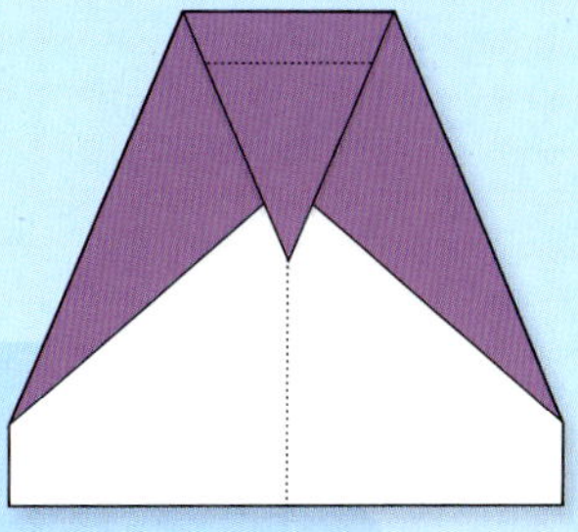

4 Fold down the tip.

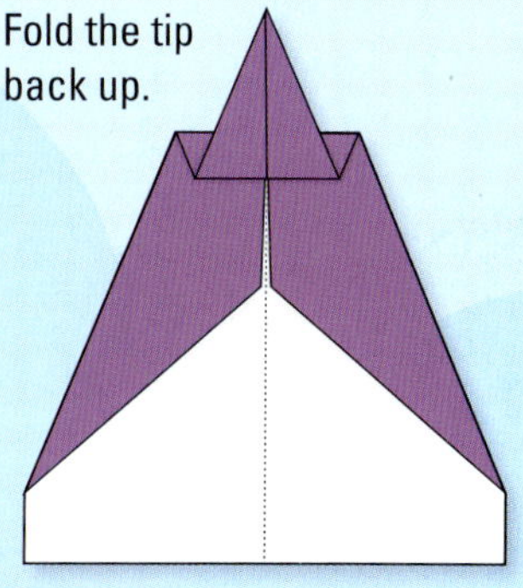

5 Fold the tip back up.

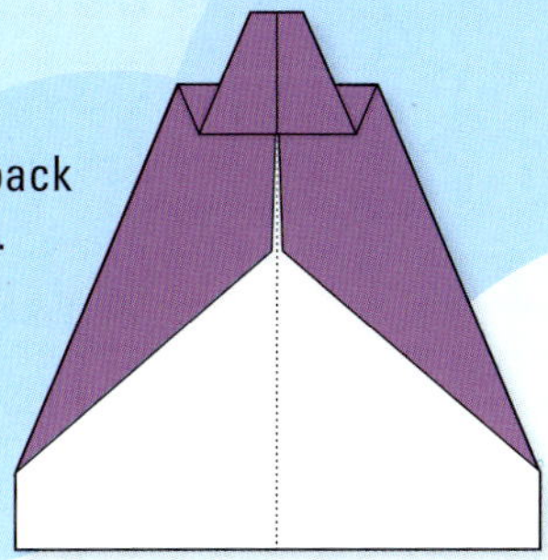

6 Fold the top tip back (away from you).

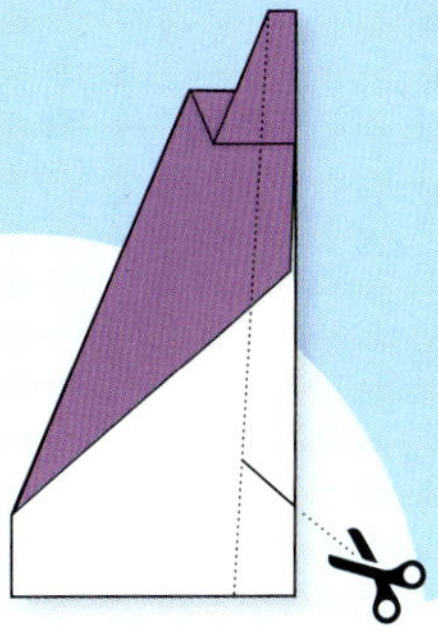

7 Fold the plane in half along the center line, with the last fold on the inside. Cut the tail along the solid cutting line. Then fold down each wing and pull up the tail.

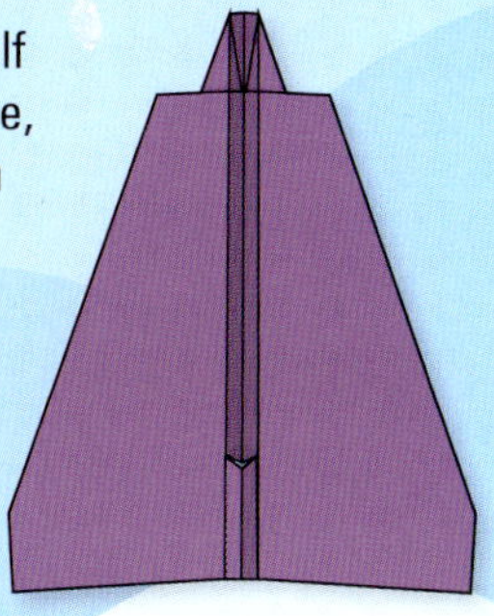

Blue Yonder

EASY | MEDIUM | HARD

The Blue Yonder is a glider that lives up to its name. With a good launch, this far-flying craft will go sailing into the blue!

1

Lay the paper flat with the name facing **down**.

Fold paper in half (from left to right).

2

Fold back each top corner.

3 Fold the top point down to create a crease.

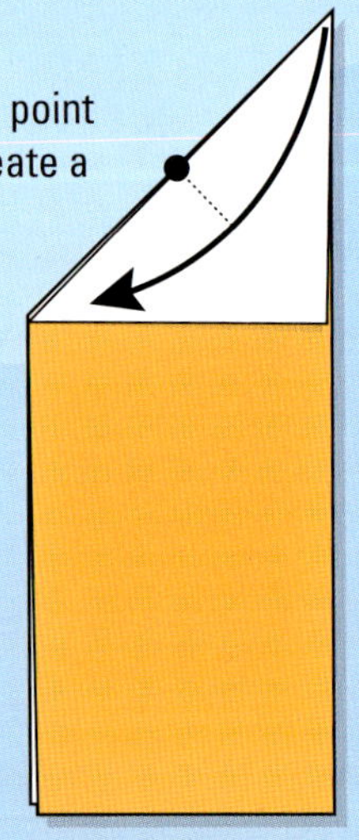

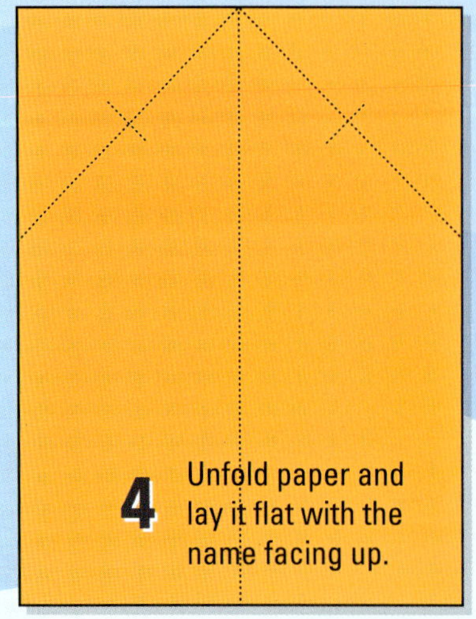

4 Unfold paper and lay it flat with the name facing up.

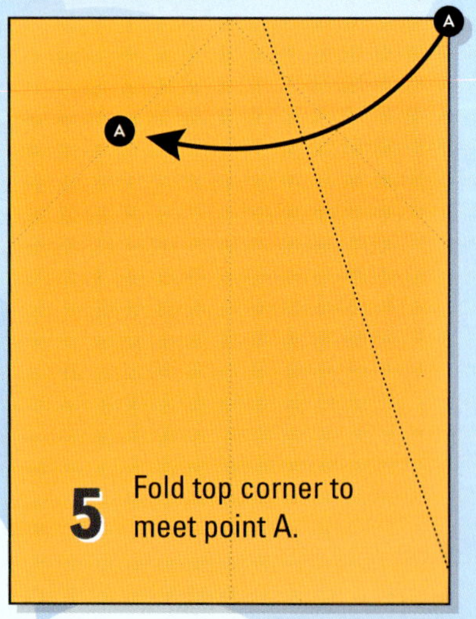

5 Fold top corner to meet point A.

6

Unfold and repeat on the other side.

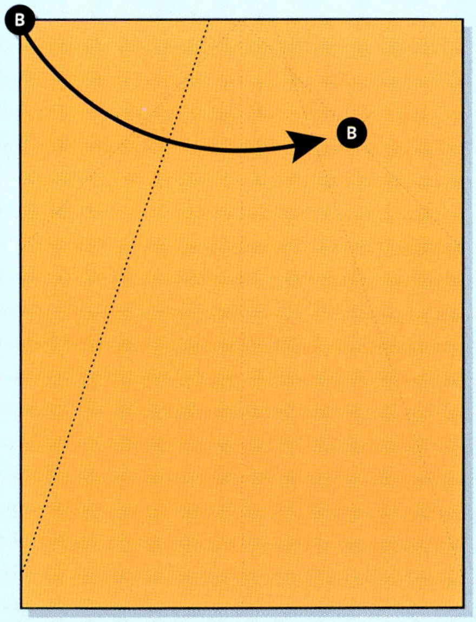

7

Fold the two sides along the creases you just made. Fold and tuck inward the two points.

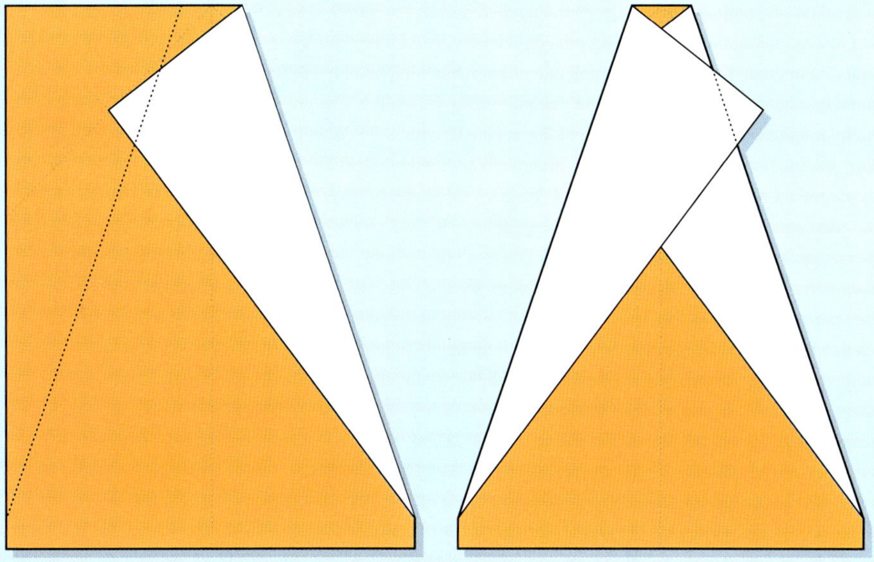

8

Fold down the top.

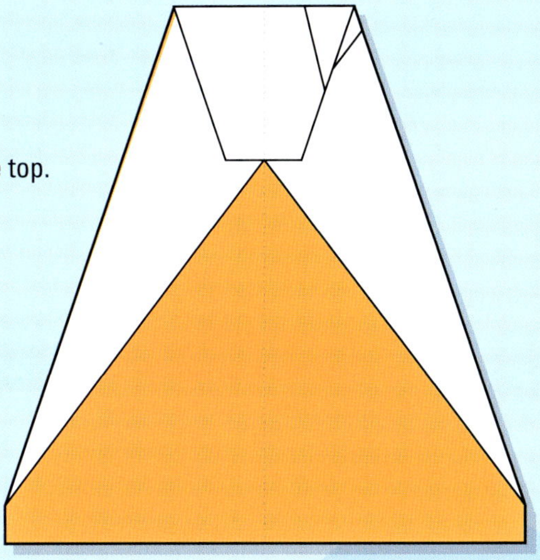

9

Fold the plane in half. Make sure you fold it so that the fold in step 8 is on the **outside**.

10

Fold down each top corner.

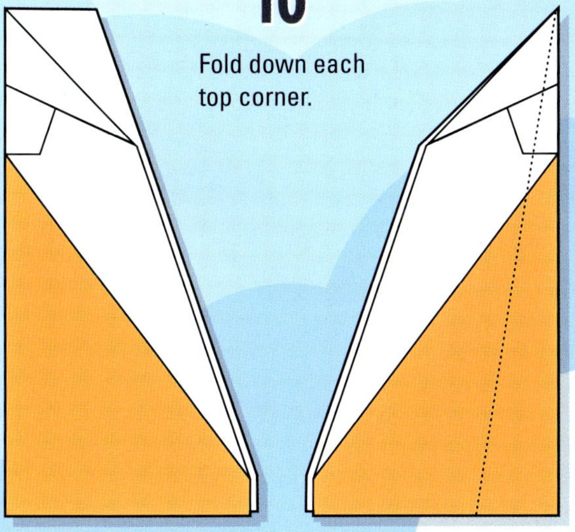

11

Fold down each wing.

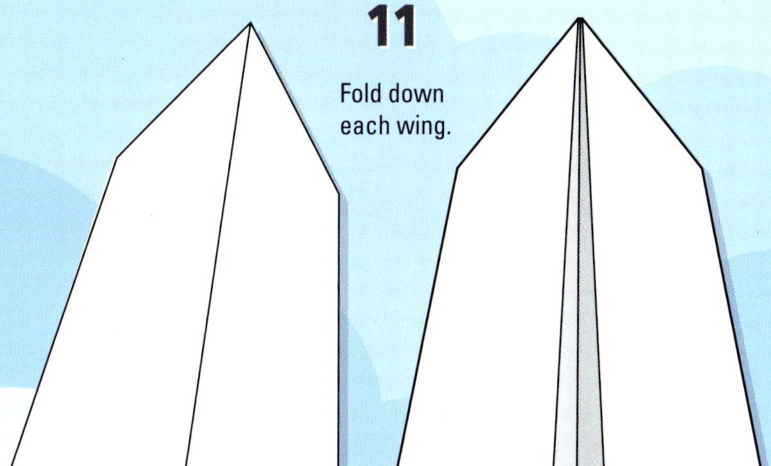

GRASSHOPPER

The Grasshopper has a shape reminiscent of classic bush planes made by Cessna and Piper. It's the perfect light craft for flitting here and there. Add a little tape to hold the wings and tail together.

EASY | MEDIUM | HARD

Lay the paper flat with the name facing **down**.

1

Fold paper in half on center line, and open up the paper so it lies flat.

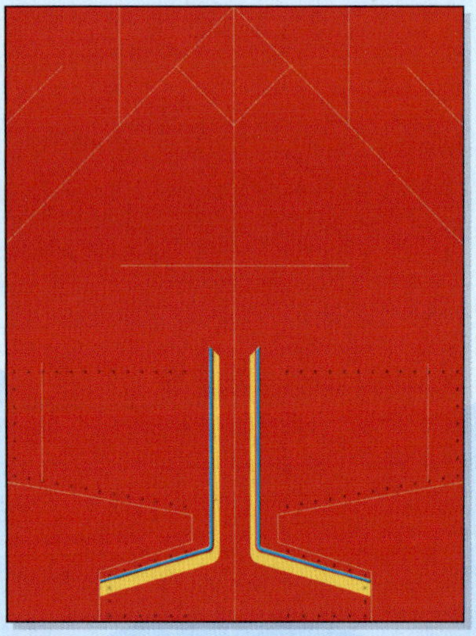

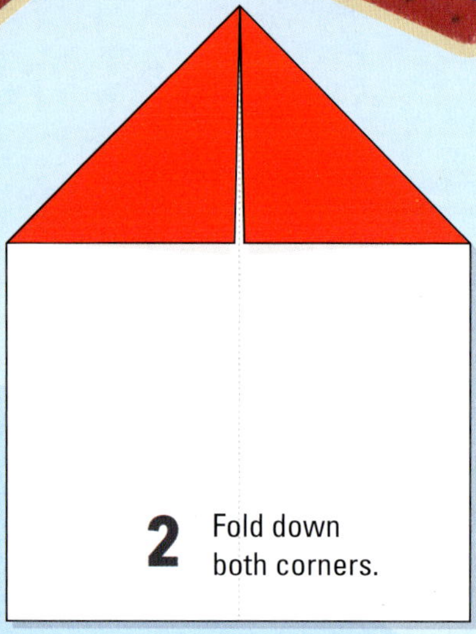

2 Fold down both corners.

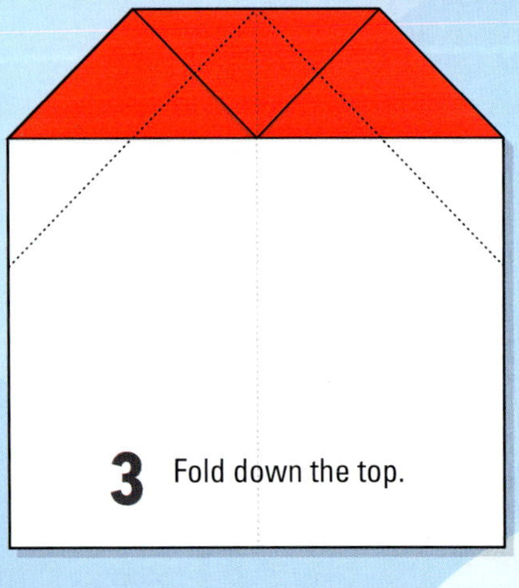

3 Fold down the top.

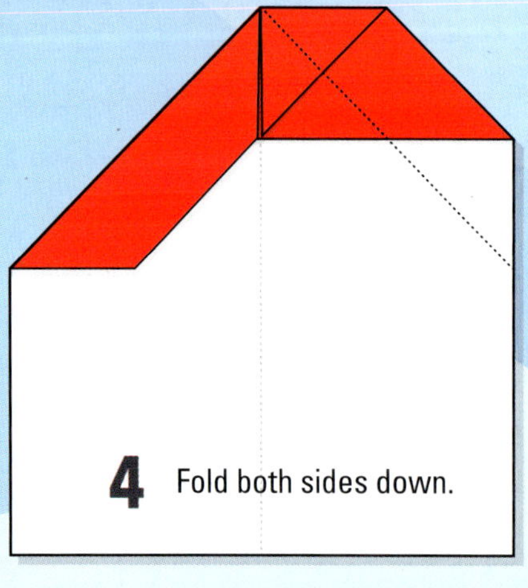

4 Fold both sides down.

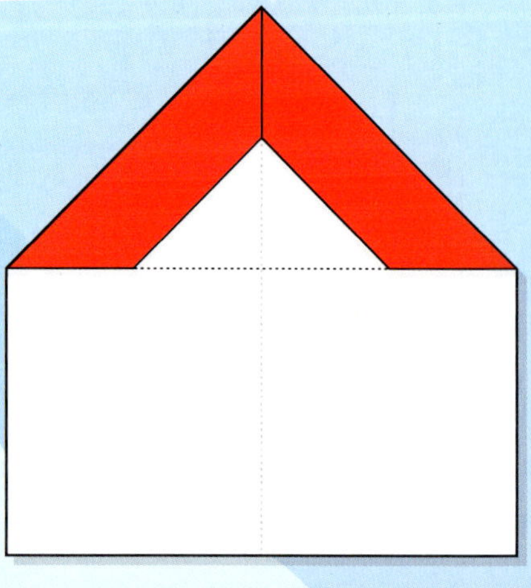

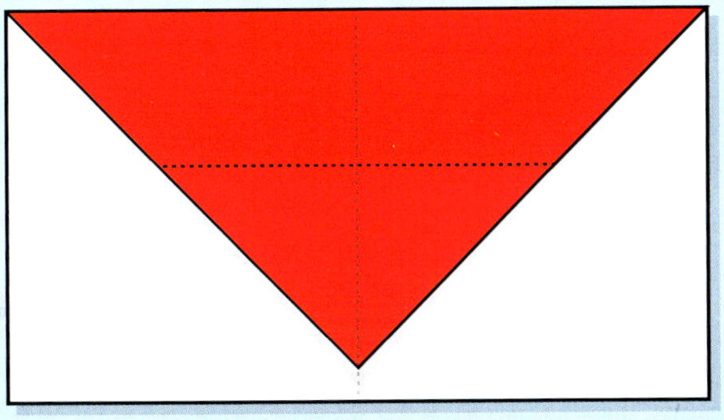

5 Fold down the tip.

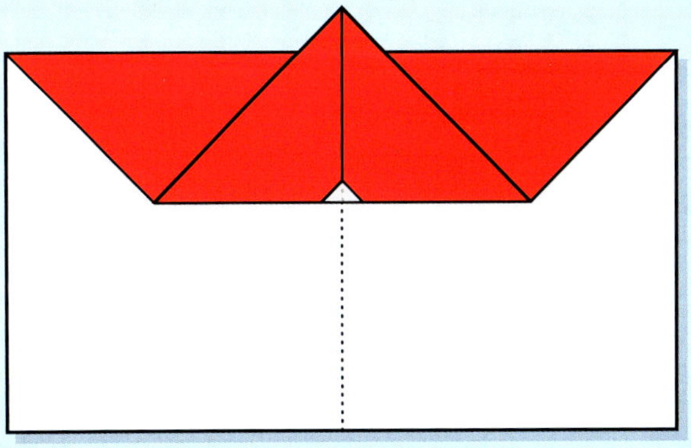

6 Fold the tip back up.

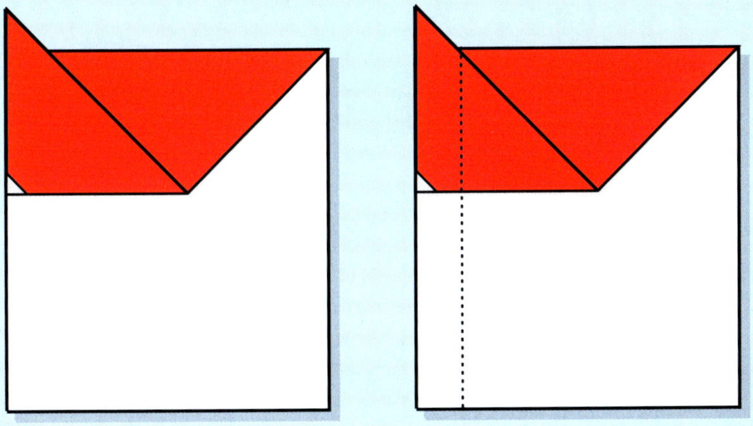

7 Fold the plane in half. Make sure you fold it so the fold in step 6 is on the outside.

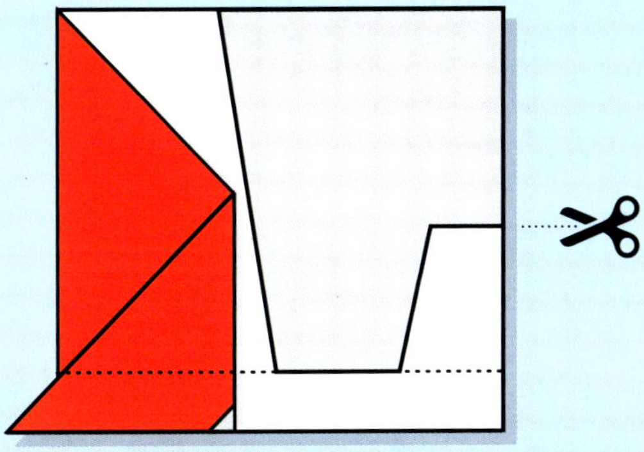

8 Cut out the shaded piece along the solid cutting line. Then fold down each wing.

9 Fold up the winglets.

FLEUR DE LOOP

EASY | MEDIUM | HARD

The Fleur de Loop is a natural acrobat capable of soaring, looping, and diving. Try adjusting its wing flaps or taping its nose to change its flight behavior.

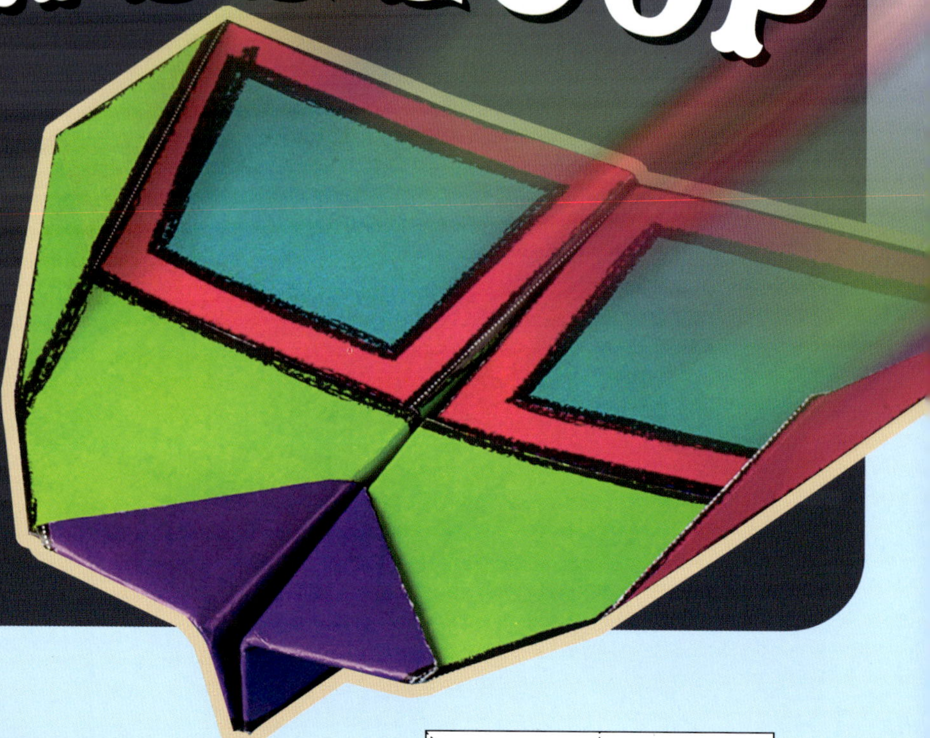

Lay the paper flat with the name facing **down**.

1 Fold paper in half on center line, and open up the paper so it lies flat.

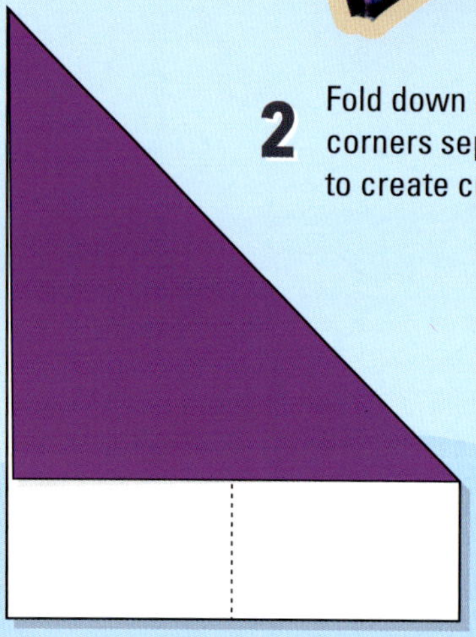

2 Fold down both corners separately to create creases.

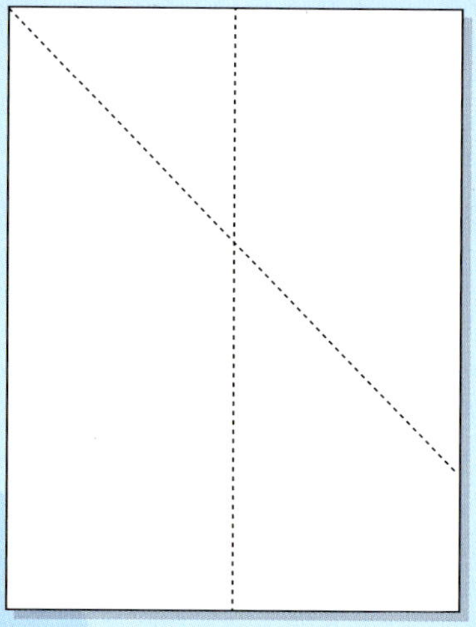

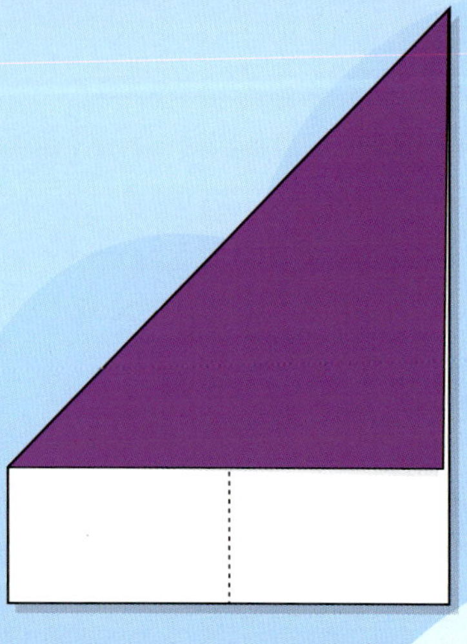

3 Fold down the top to make another crease. Then lay the paper flat.

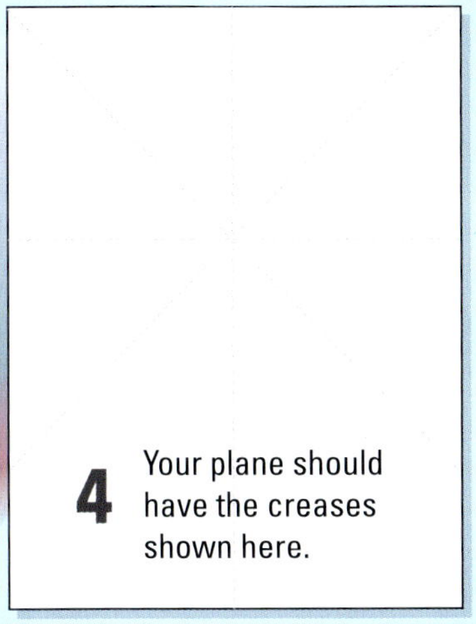

4 Your plane should have the creases shown here.

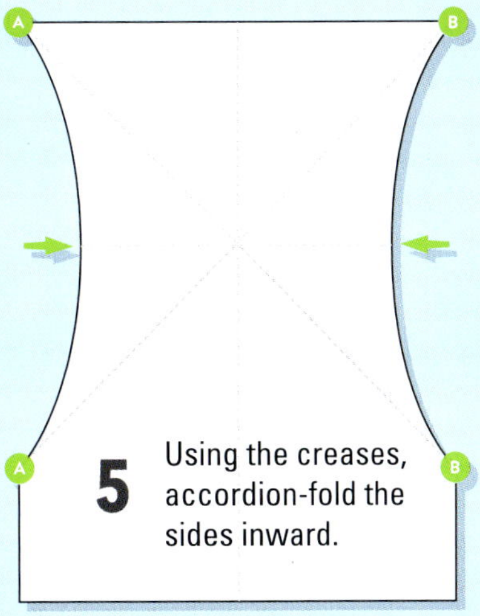

5 Using the creases, accordion-fold the sides inward.

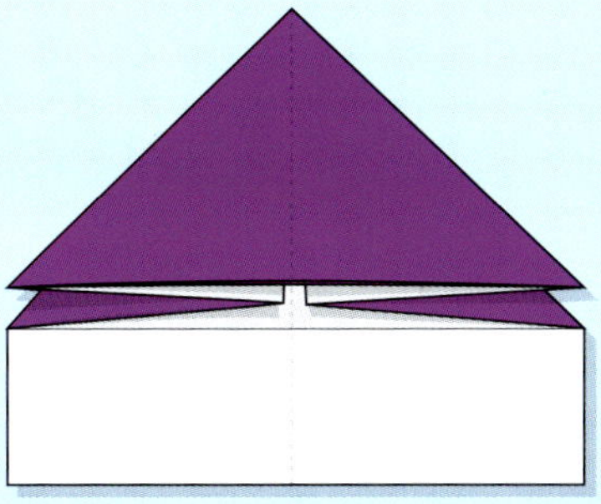

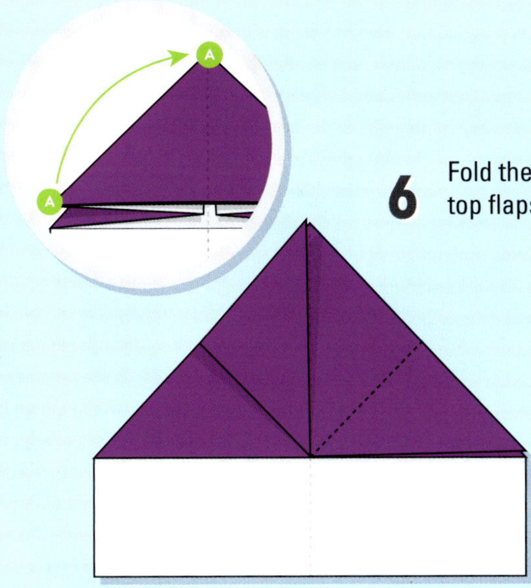

6 Fold the two top flaps up.

7 Fold the top of each flap down.

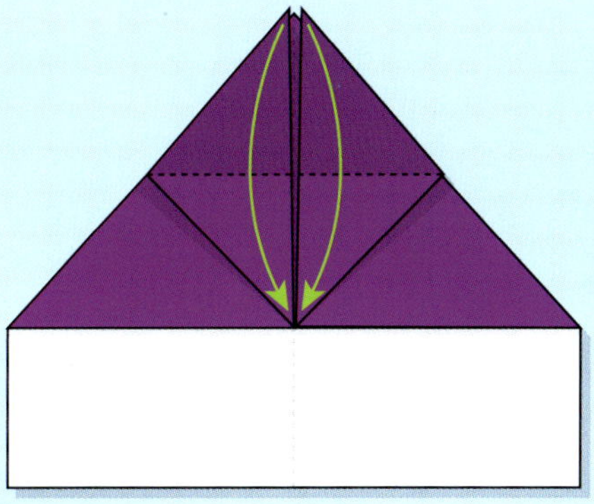

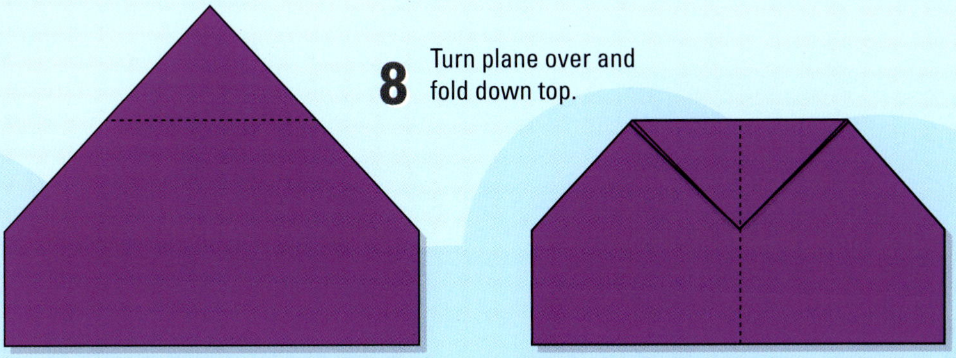

8 Turn plane over and fold down top.

9 Fold plane in half. Make sure you fold it so that the fold in step 7 is on the outside.

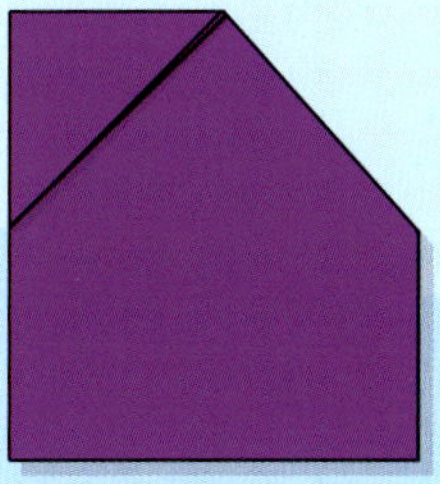

10 Fold down each wing.

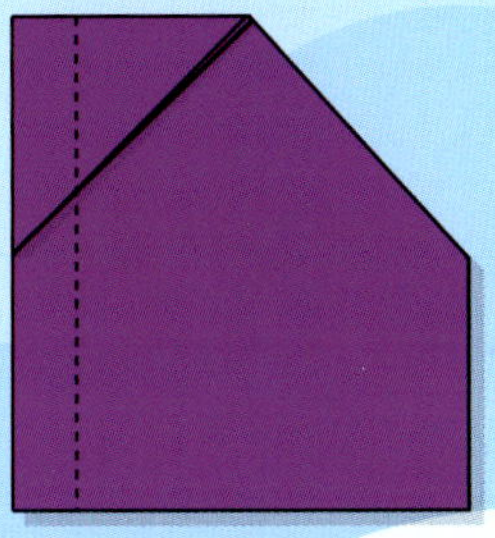

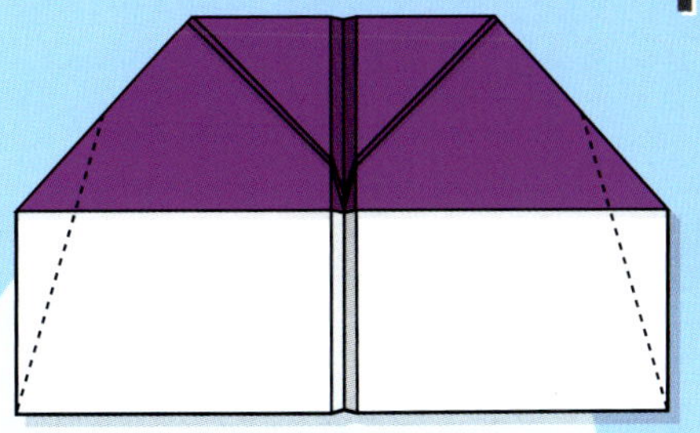

11 Fold up the winglets.

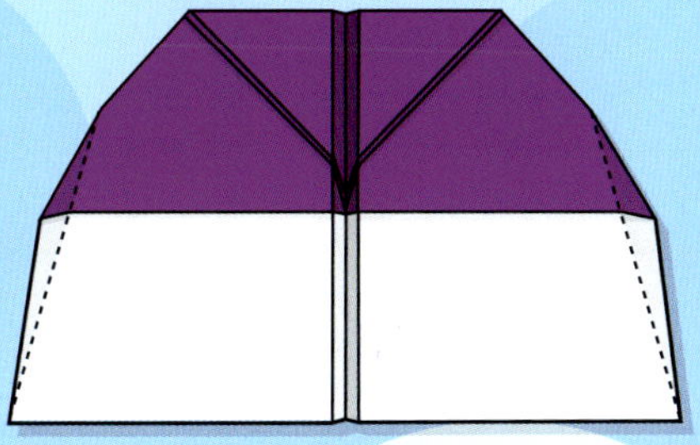

Jolly Dodger

EASY | MEDIUM | HARD

Make sure the forward flaps are well-creased or you'll find this dodgy vessel veering off. If it pitches upward too much, you can straighten out its course by adding one or two paper clips to the nose.

Lay the paper flat with the name facing **up**.

1 Fold paper in half on center line, and open up the paper so it lies flat.

2 Fold down the top.

3 Turn the plane over. Fold down both sides.

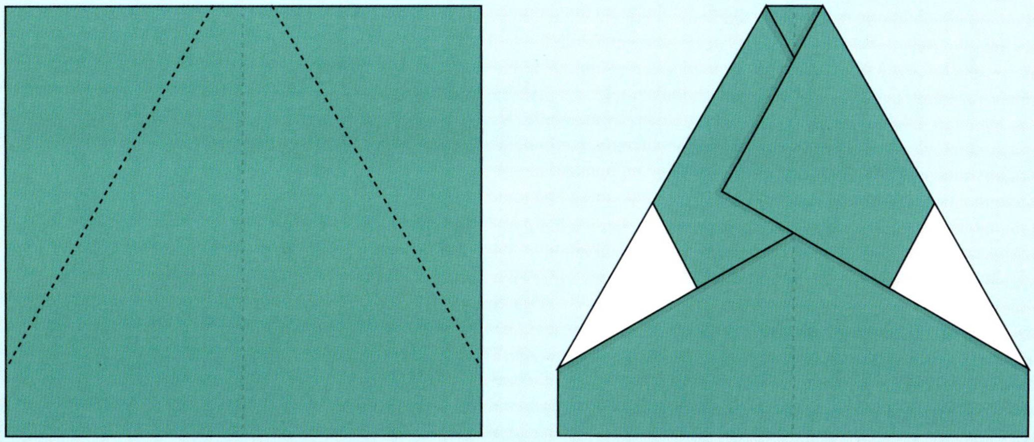

4 Fold each side back outward.

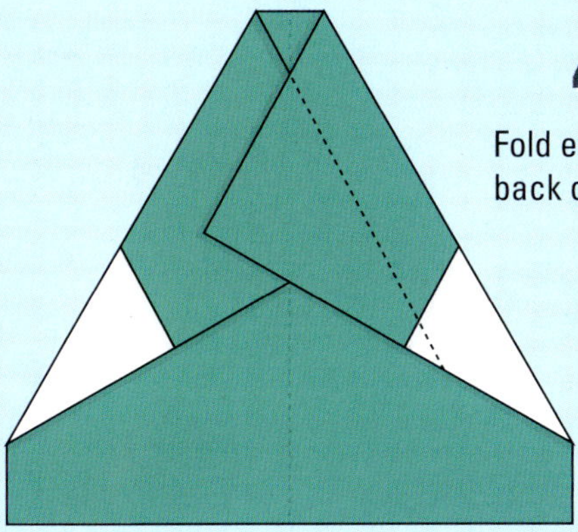

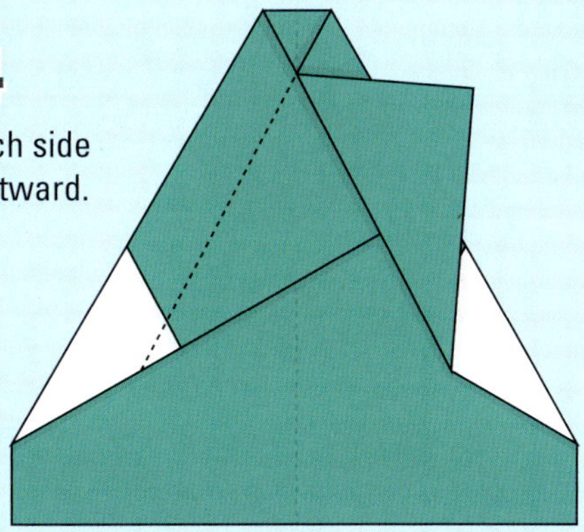

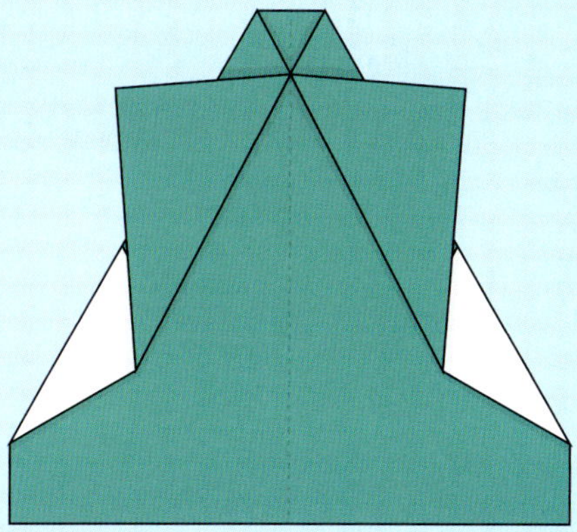

5 Fold the plane in half along the center line, with the last fold on the inside. Then fold down each wing.

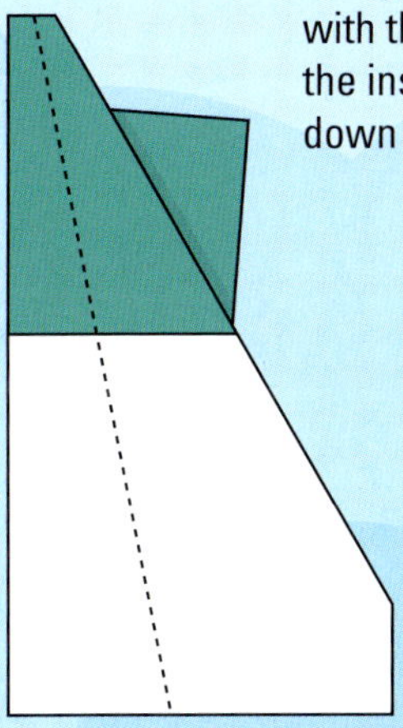

6 Fold up the winglets.

Swerving Dervish

EASY | MEDIUM | HARD

A wild look for a wild ride! The Swerving Dervish has a swoop-and-stall flight pattern that makes for an unpredictable landing. Follow the lines as closely as you can when folding this plane; it has an intricate fold in step 8 that requires precision.

Lay the paper flat with the name facing **up**.

1 Fold the right corner down so that it makes this shape. Cut the bottom piece off along the solid cutting line. SAVE this piece. Unfold and make the same fold with the left corner.

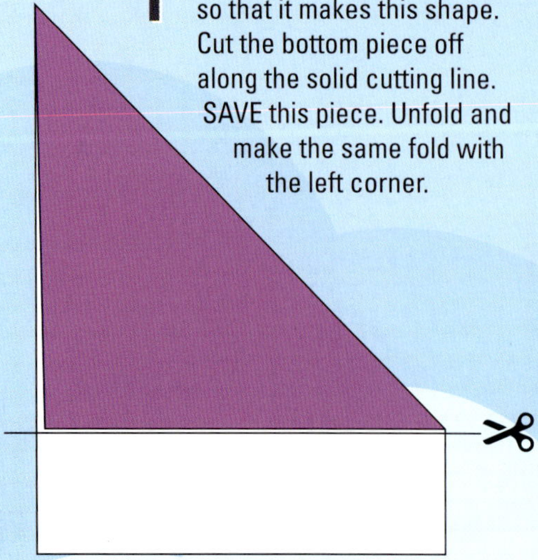

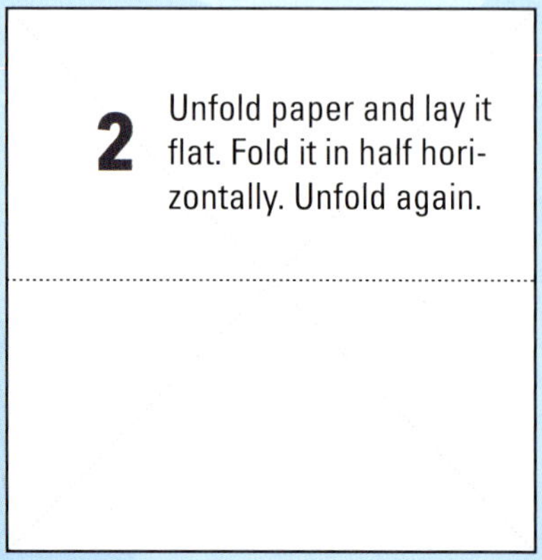

2 Unfold paper and lay it flat. Fold it in half horizontally. Unfold again.

3 Using the creases, accordion-fold the sides inward to create a triangular shape. Match points A to A and B to B while folding.

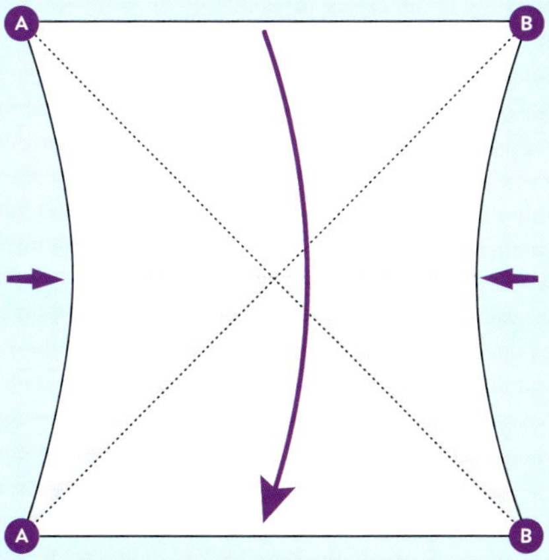

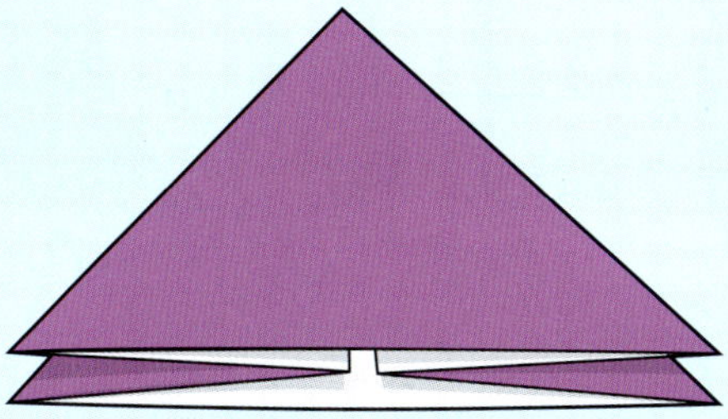

4 Fold both top flaps up.

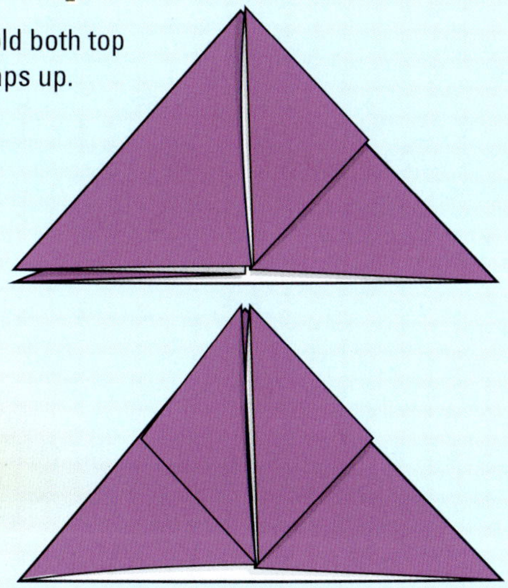

5 Fold down both top corners of the piece of paper cut off in step 1.

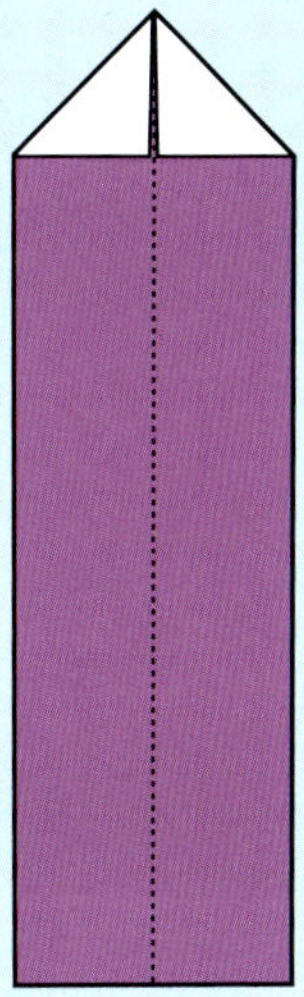

6 Insert this paper into the plane.

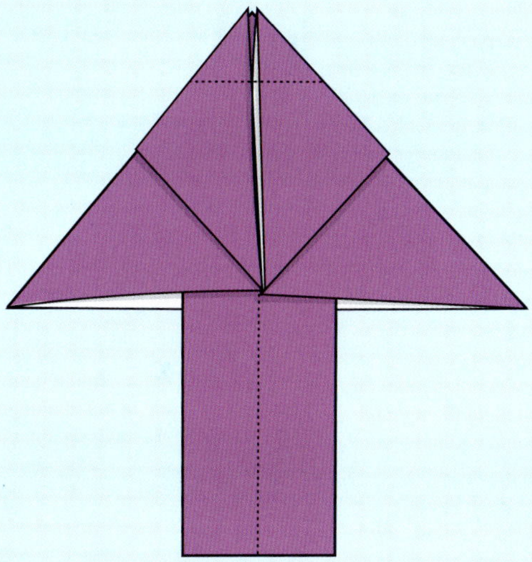

7 Fold down the top to create a crease.

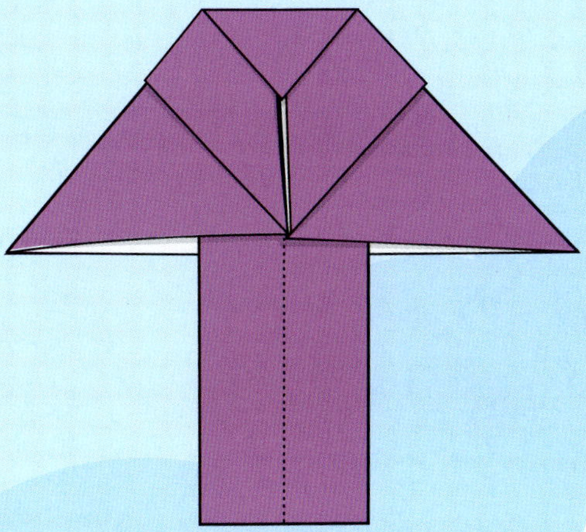

8 Tuck the inner flaps into the slits of the outer flap.

9 Fold in half.

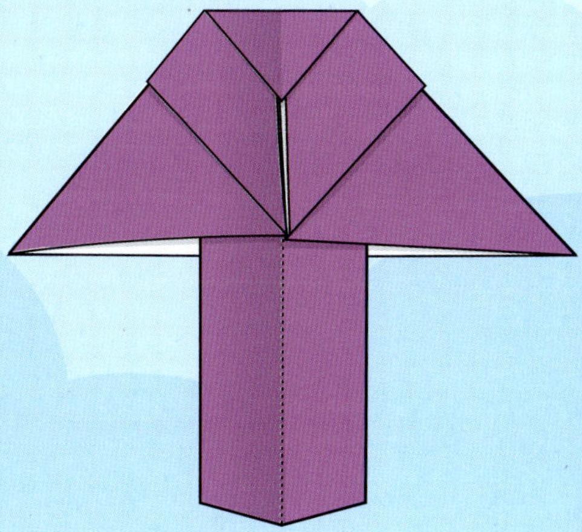

BIG MOUTH

The odd tubular front actually stabilizes this model's flight. There are some difficult steps involved, so make sure you take your time and line up the folds correctly.

EASY MEDIUM HARD

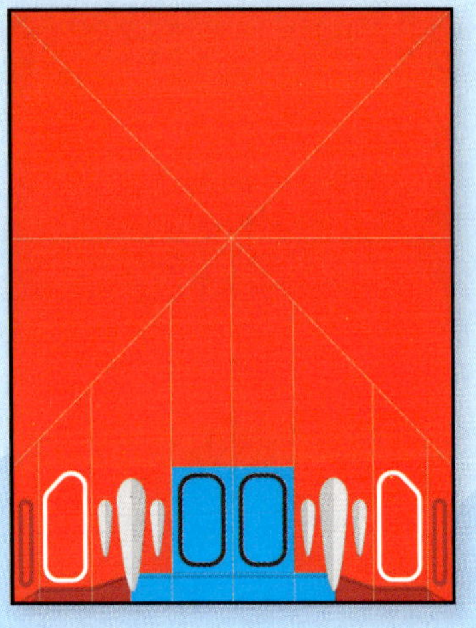

Lay the paper flat with the name facing **down**.

1 Fold the left and right corners and then lay it flat again.

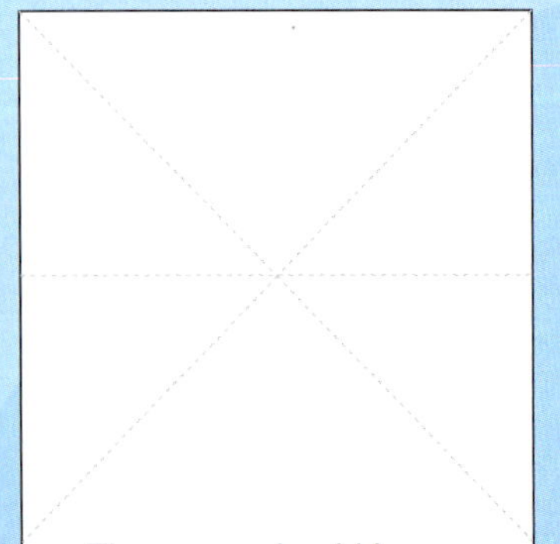

The paper should have these creases.

2 Fold down the top away from you to make another crease. Then lay the paper flat.

3 Using the creases, accordion-fold the sides inward. Match points A to A and B to B while folding.

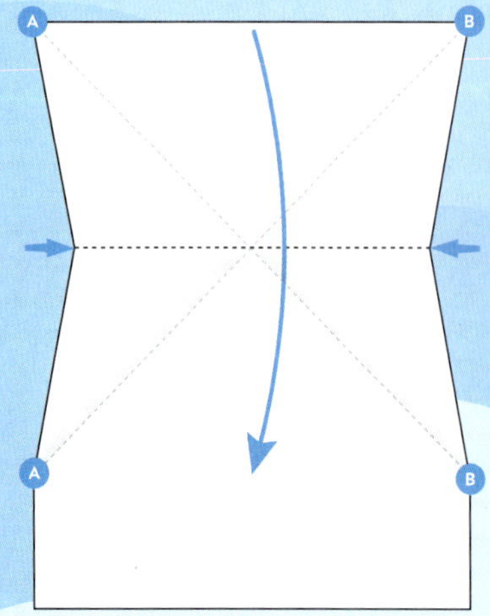

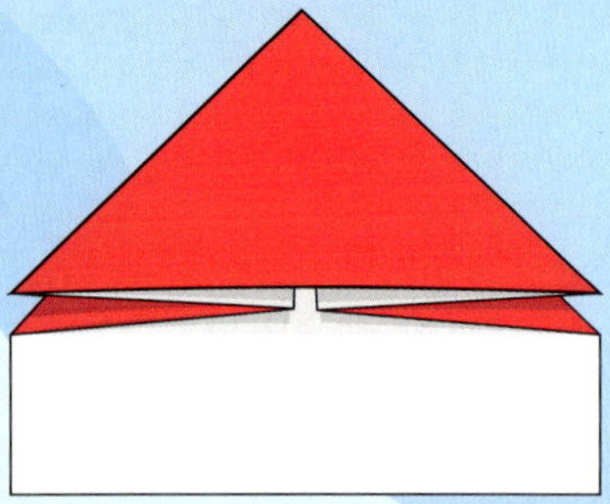

5 Tuck one corner into the other. Wiggle them back and forth to make the fit snug.

4 Fold each side flap inward.

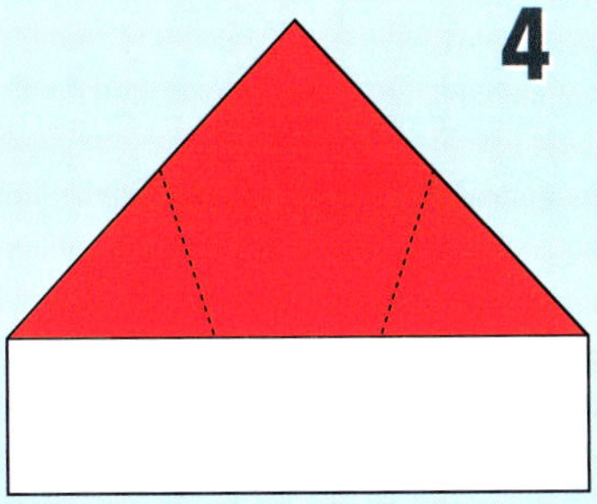

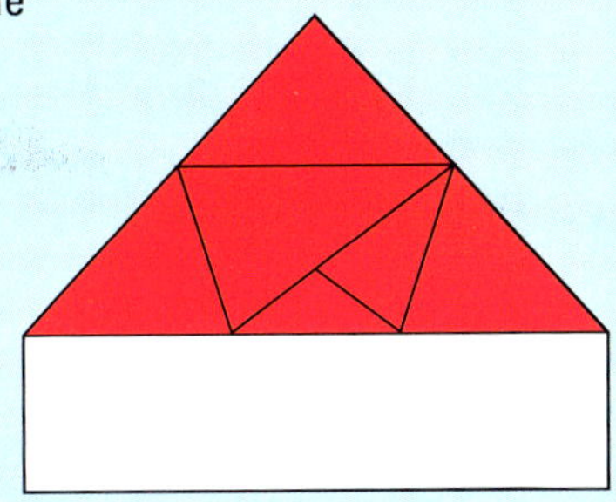

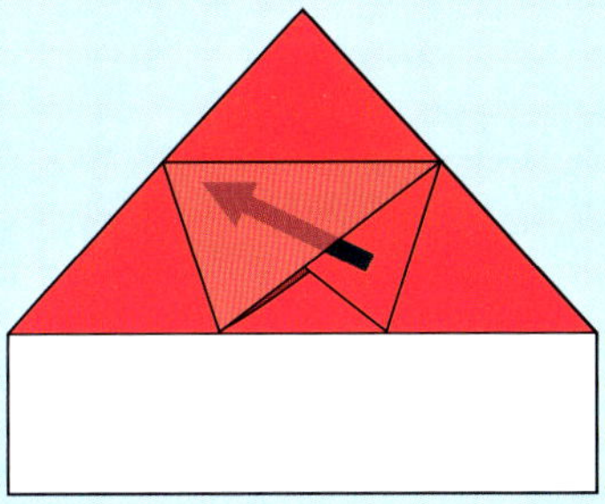

6 Turn the plane over and fold down the top.

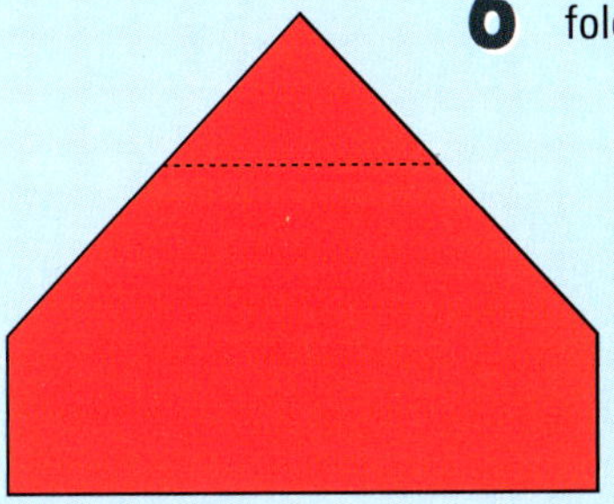

7 Turn the plane over and fold the plane along the center line.

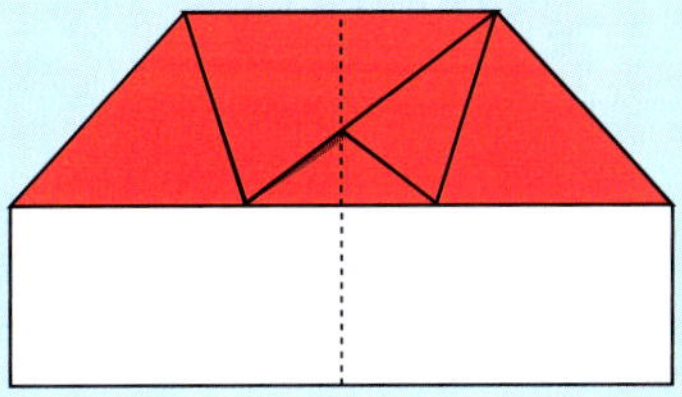

8 Use the crease to help you open up the mouth, as shown here.

9 Pinch the top of the mouth. Create two creases to give the mouth a shape.

10 Fold down wings.

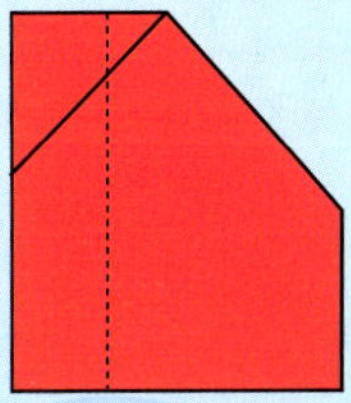

11 Fold down each wing's leading edge and tuck it inside the underflap of the nose.

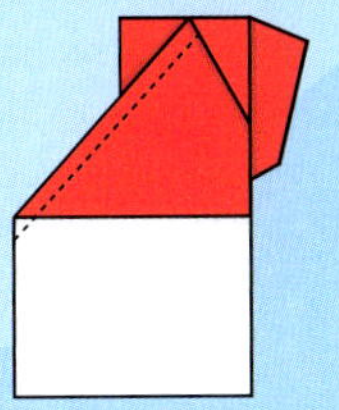

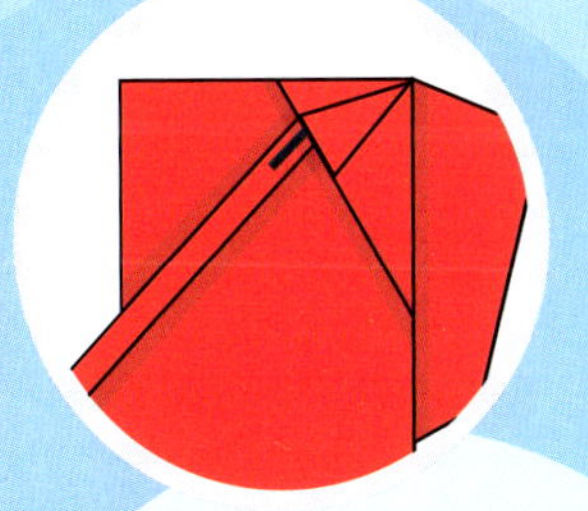

12 Fold up each pair of winglets.

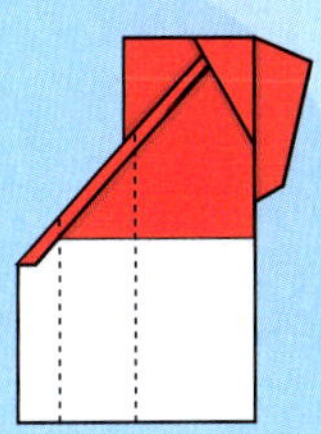

STAR SPRITE

A sleek retro craft ready for the stars! The Star Sprite is one of the trickiest models to build–you'll need to fold very precisely–but the result is a slender and nimble spaceship perfectly suited for surfing alien moonscapes.

EASY MEDIUM HARD

Lay the paper flat with the name facing **down**.

1 Fold paper in half on center line and open up the paper so it lies flat.

2 Fold down both corners.

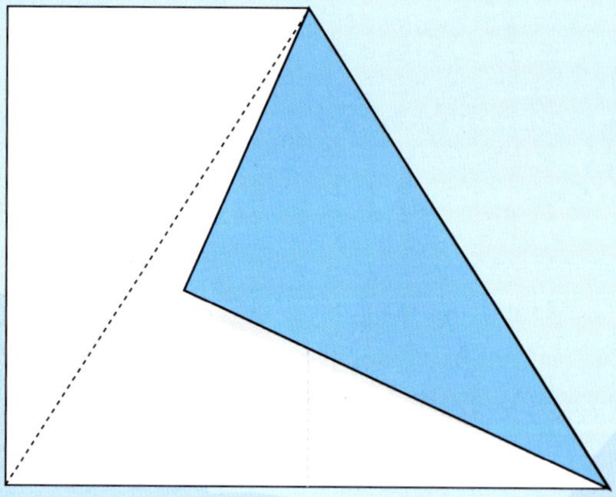

3 Fold both sides in to the center line.

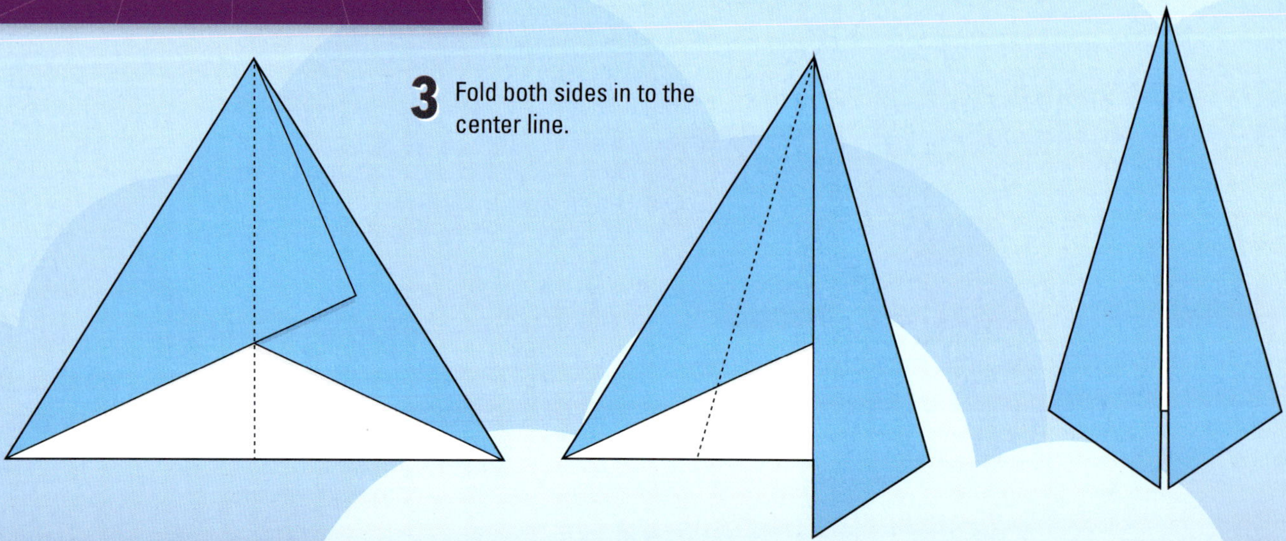

4 Fold lower corners back.

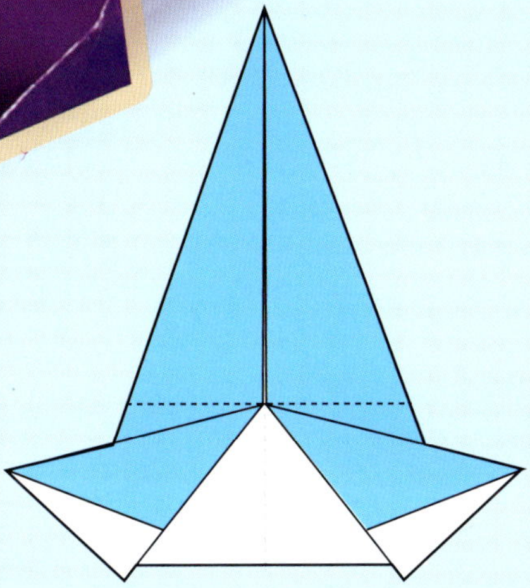

5 Fold the top over. Then fold back up along the second line.

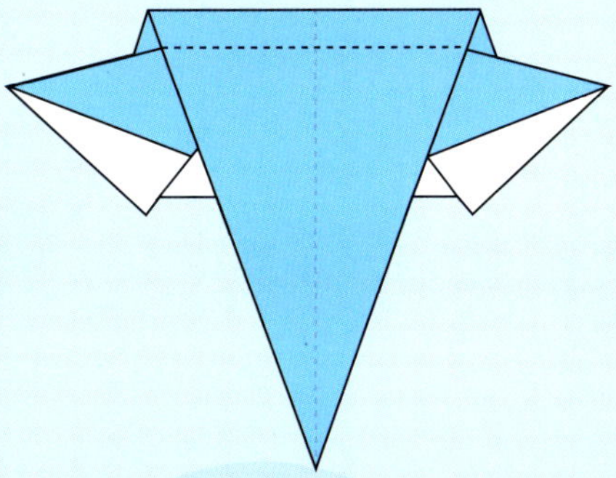

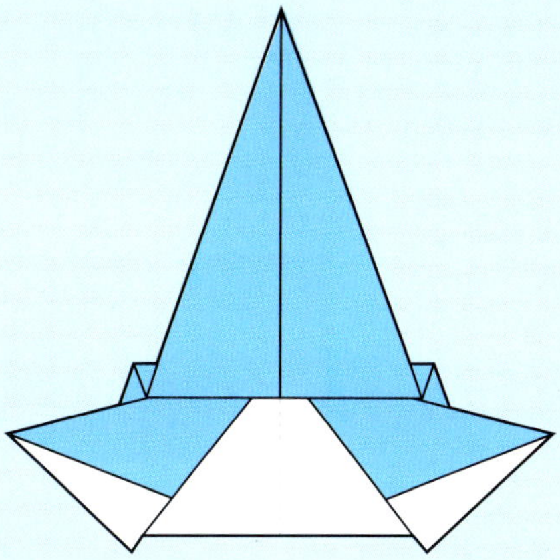

6 Fold the plane in half along the center line. Make sure you fold it so that the fold in step 5 is on the outside.

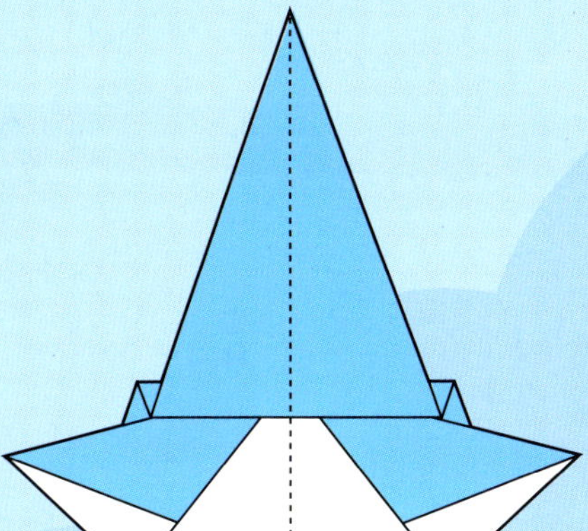

7 Fold down each wing.

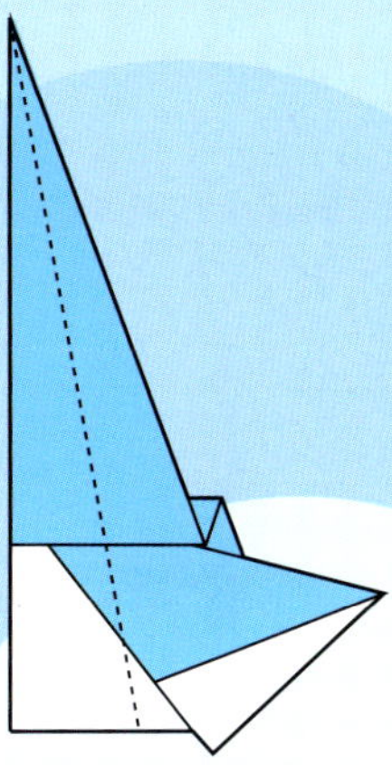

8 Fold up each pair of winglets.

Mean Streak

EASY | MEDIUM | HARD

A streak of color and it's gone–this speedy plane loves a fast launch. Experiment with the winglets to make it fly straight or do tight spirals. Make sure that you fold carefully at the beginning. This will help you line up the tricky folds in steps 5 and 6.

Lay the paper flat with the name facing **up**.

1 Fold paper in half on center line and open up the paper so it lies flat.

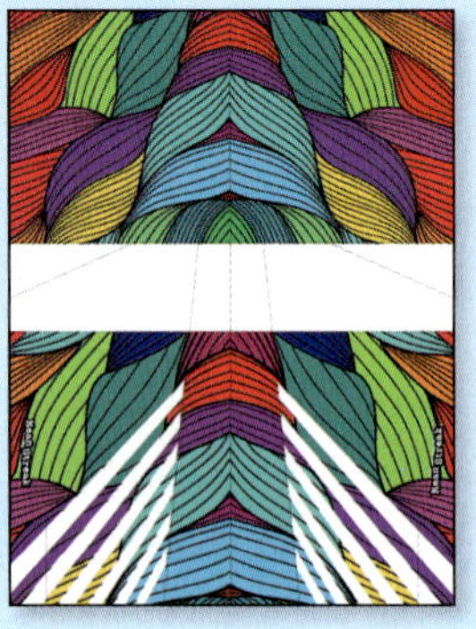

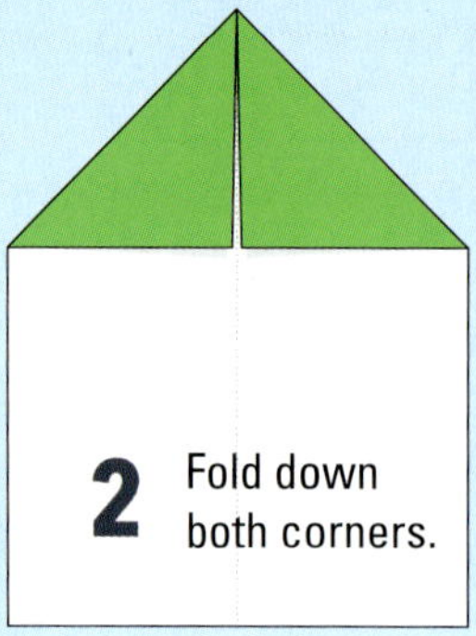

2 Fold down both corners.

3 Fold the top over. Then fold back up along the second line.

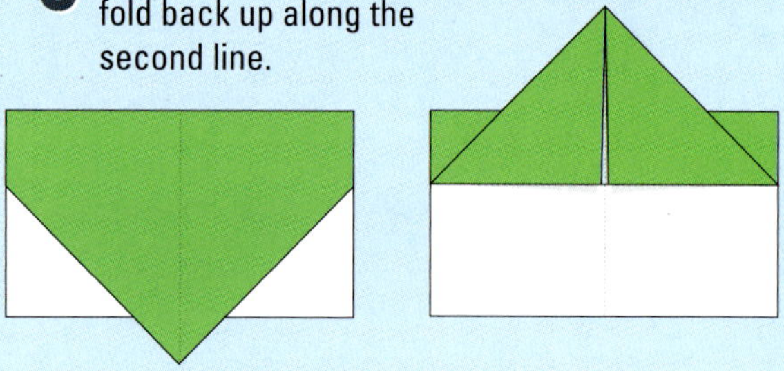

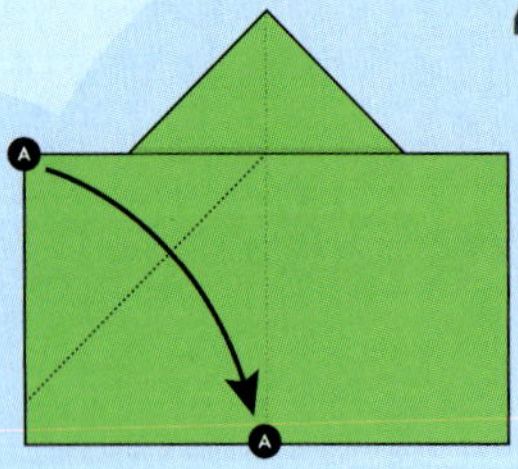

4 Turn the plane over. Then fold the top corner down (A to A). Crease this fold first.

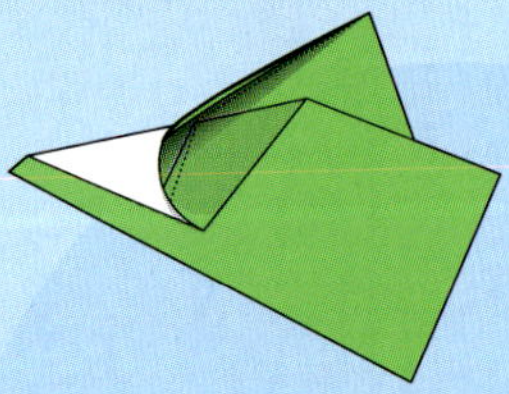

5 Make sure that the line of the top flap **runs vertically along the center line.** Press down the flaps below it (in two places) so that the plane lies flat. Repeat steps for the other side.

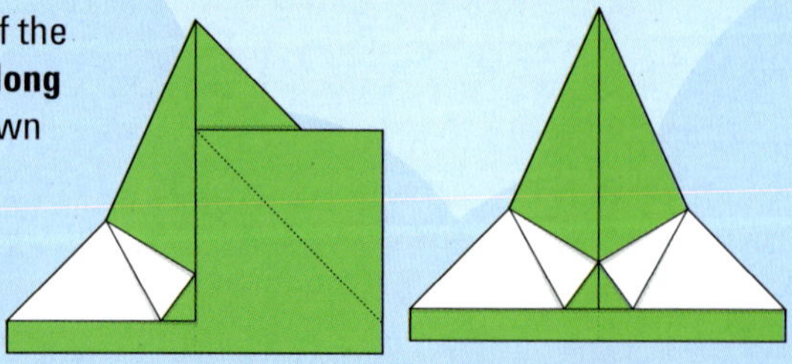

6 Fold plane in half, with the previous folds on the outside. Fold down wings.

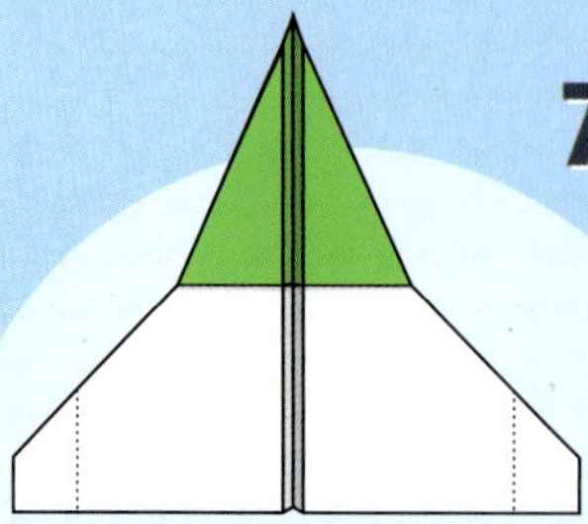

7 Fold winglets (up or down).

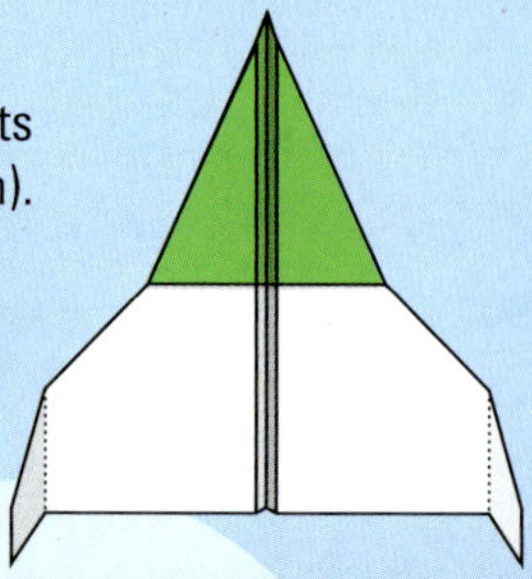

Classroom Classic
Classroom Classic

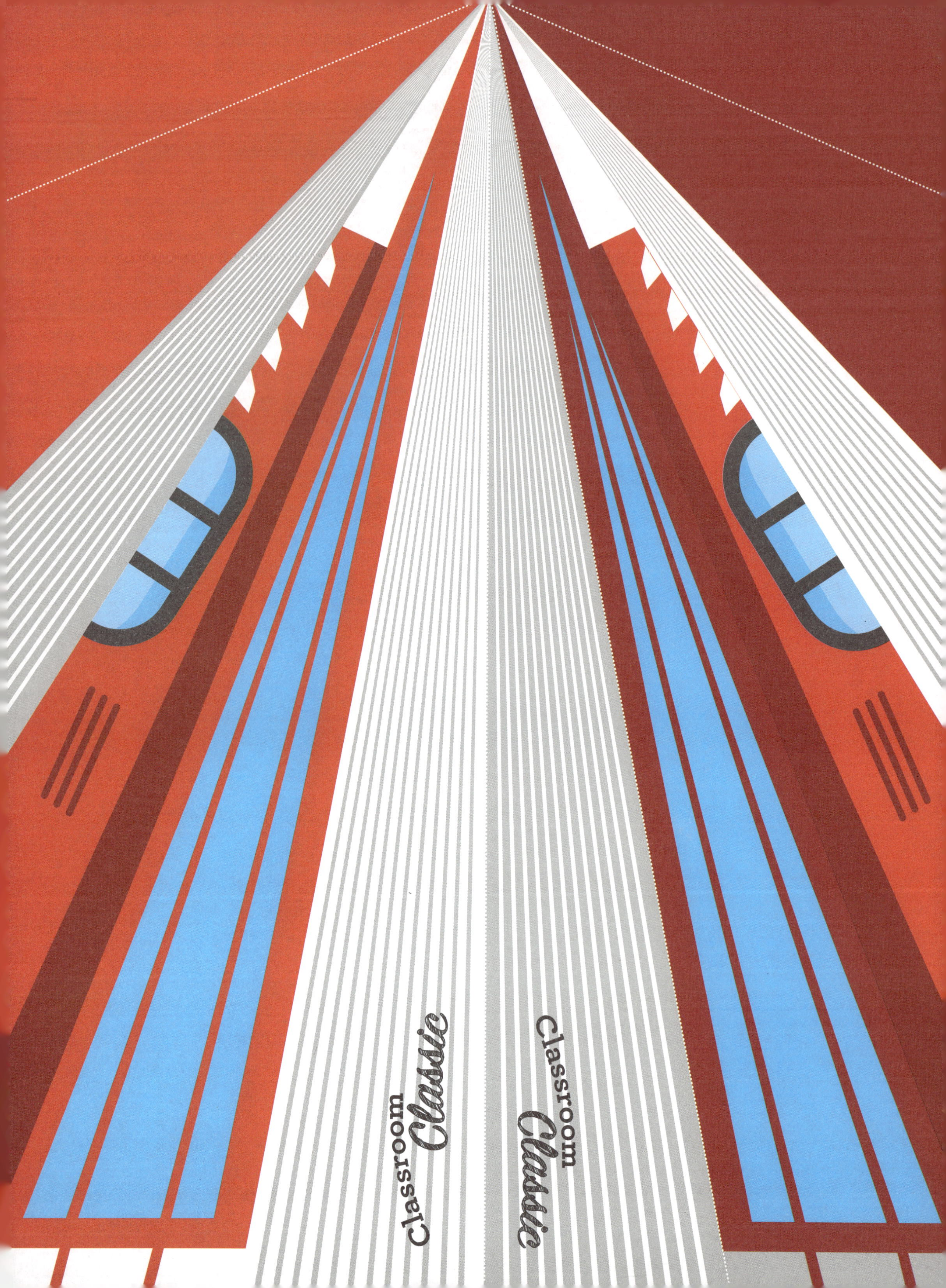
Classroom Classic
Classroom Classic

Classroom Classic
Classroom Classic

Classroom Classic

Classroom Classic

Classroom Classic
Classroom Classic

Classroom Classic
Classroom Classic

THE NEEDLE

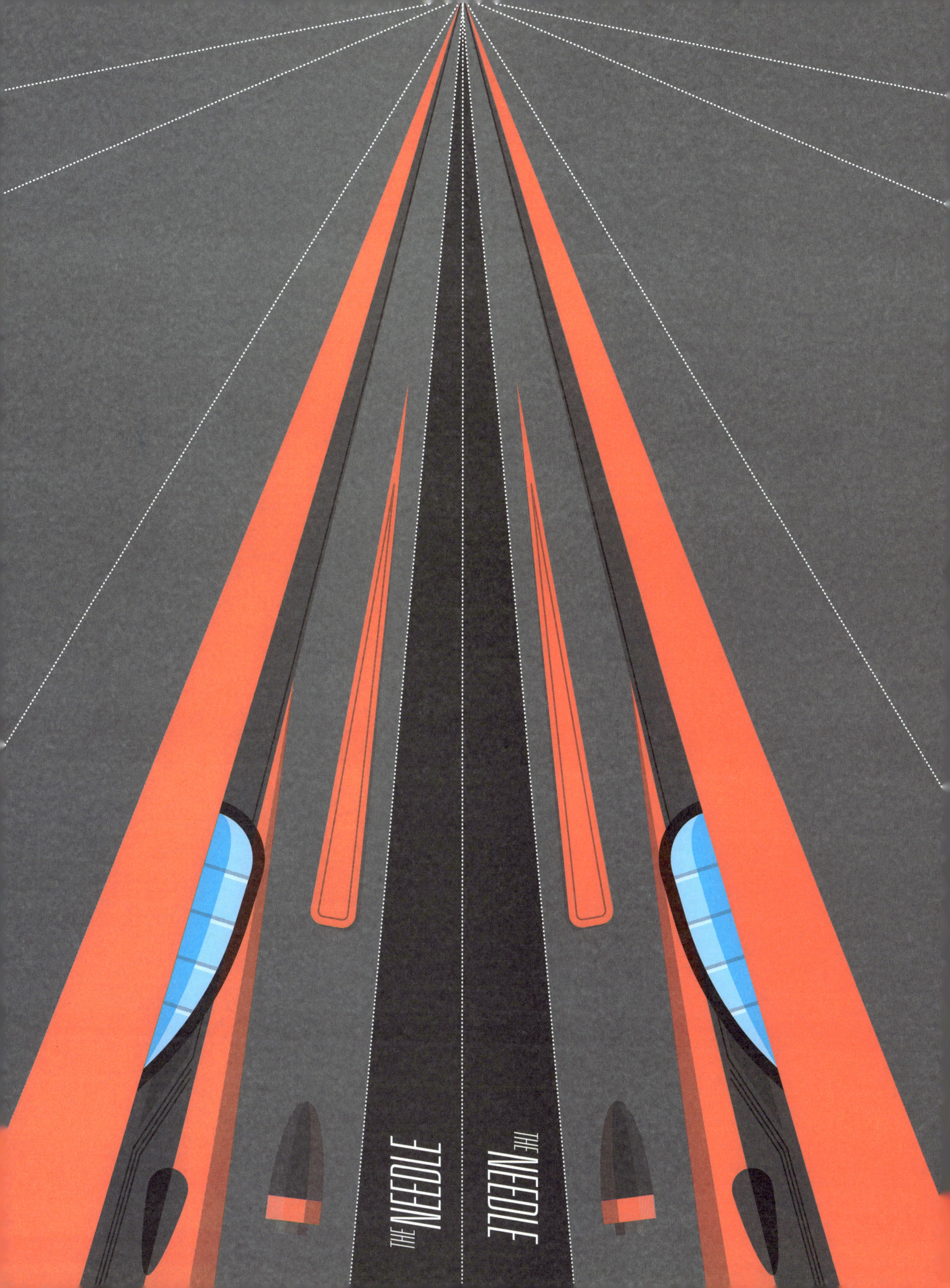
THE NEEDLE
THE NEEDLE

THE NEEDLE
THE NEEDLE

THE NEEDLE
THE NEEDLE

THE NEEDLE
THE NEEDLE

CLOUD DRAGON
CLOUD DRAGON

CLOUD DRAGON
CLOUD DRAGON

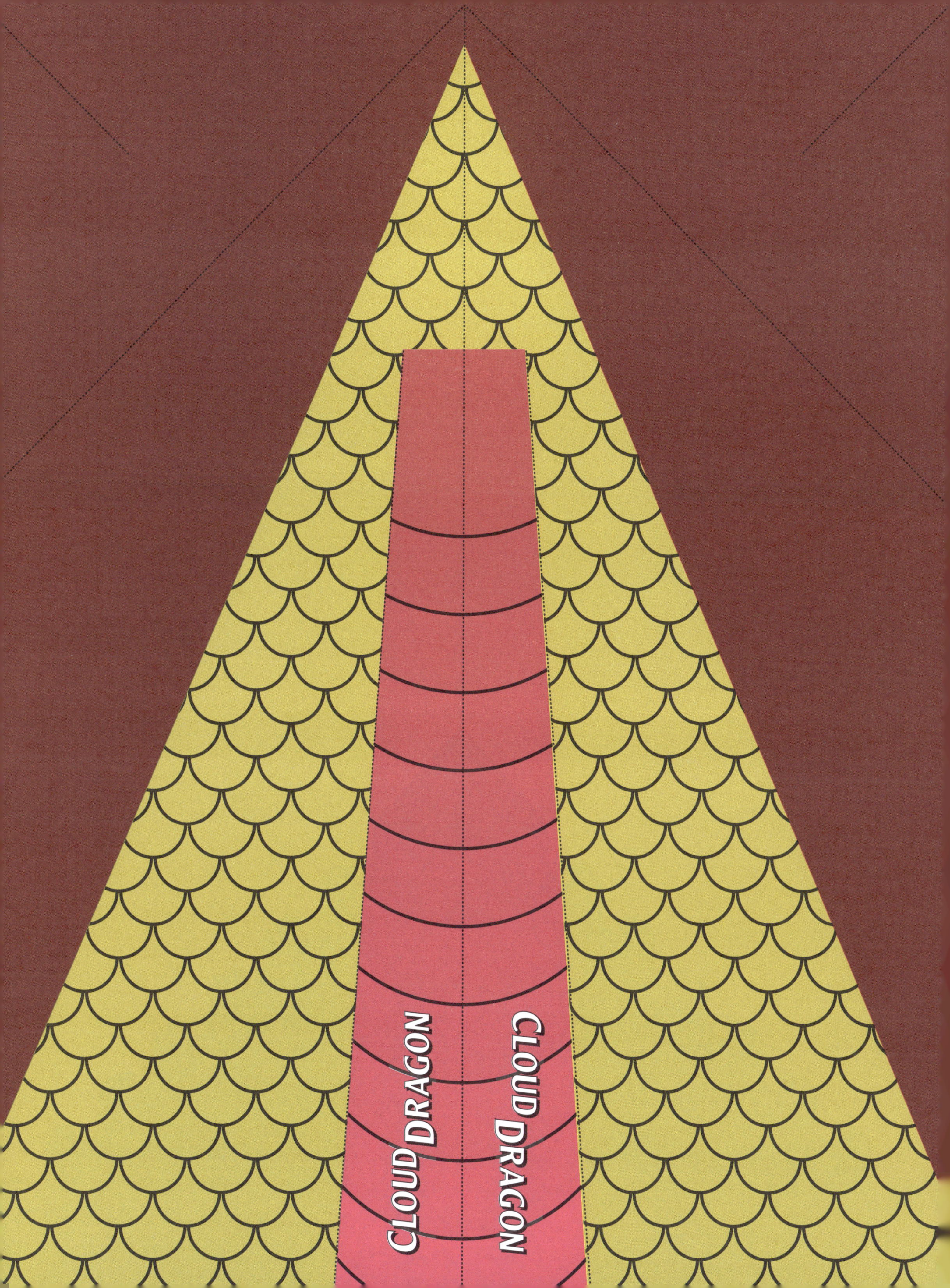
CLOUD DRAGON
CLOUD DRAGON

CLOUD DRAGON
CLOUD DRAGON

CLOUD DRAGON
CLOUD DRAGON

the fly
the fly

the fly

the fly

the fly

the fly

the fly

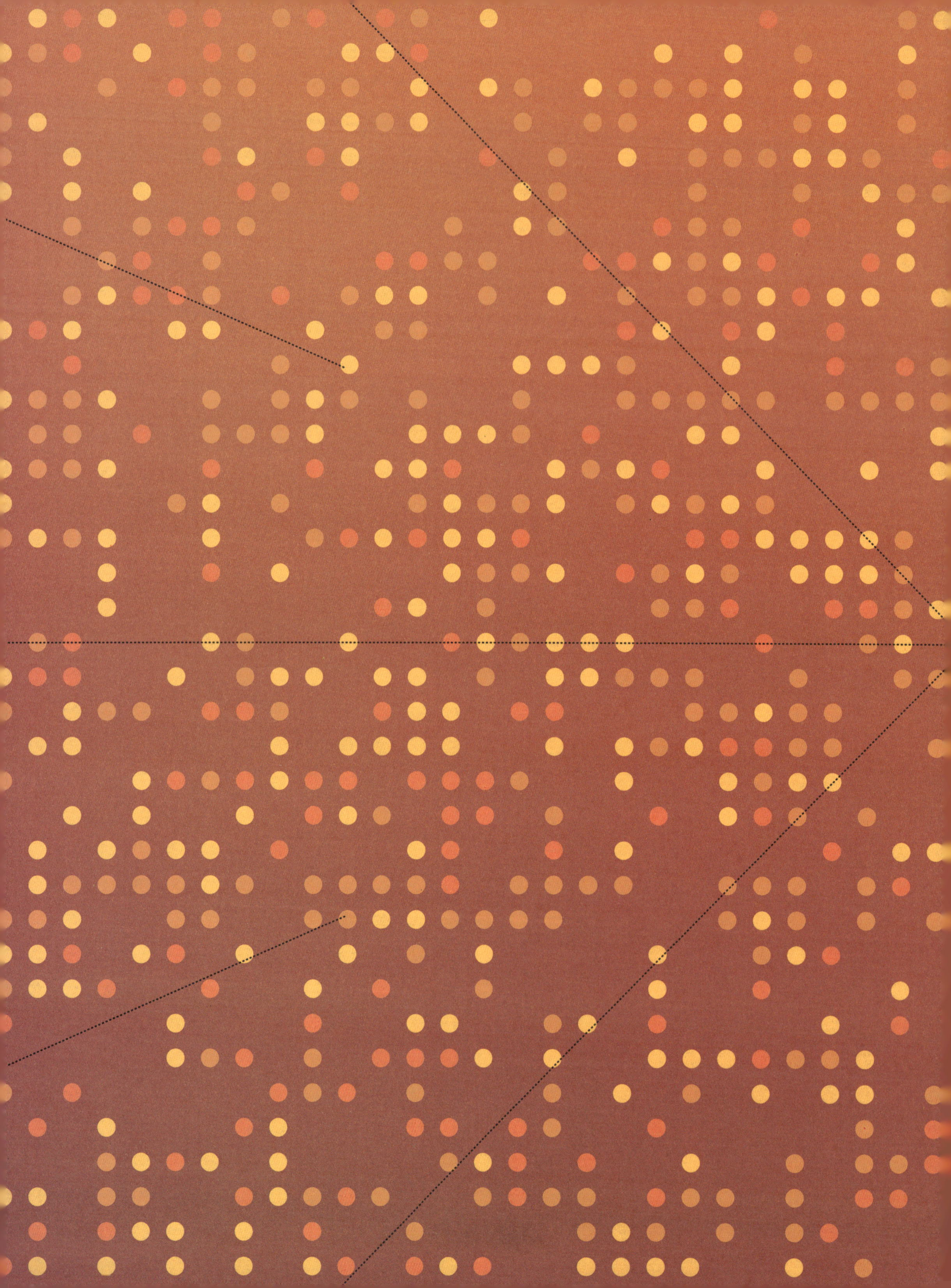

the fly

the fly

PHOENIX
PHOENIX

PHOENIX
PHOENIX

PHOENIX

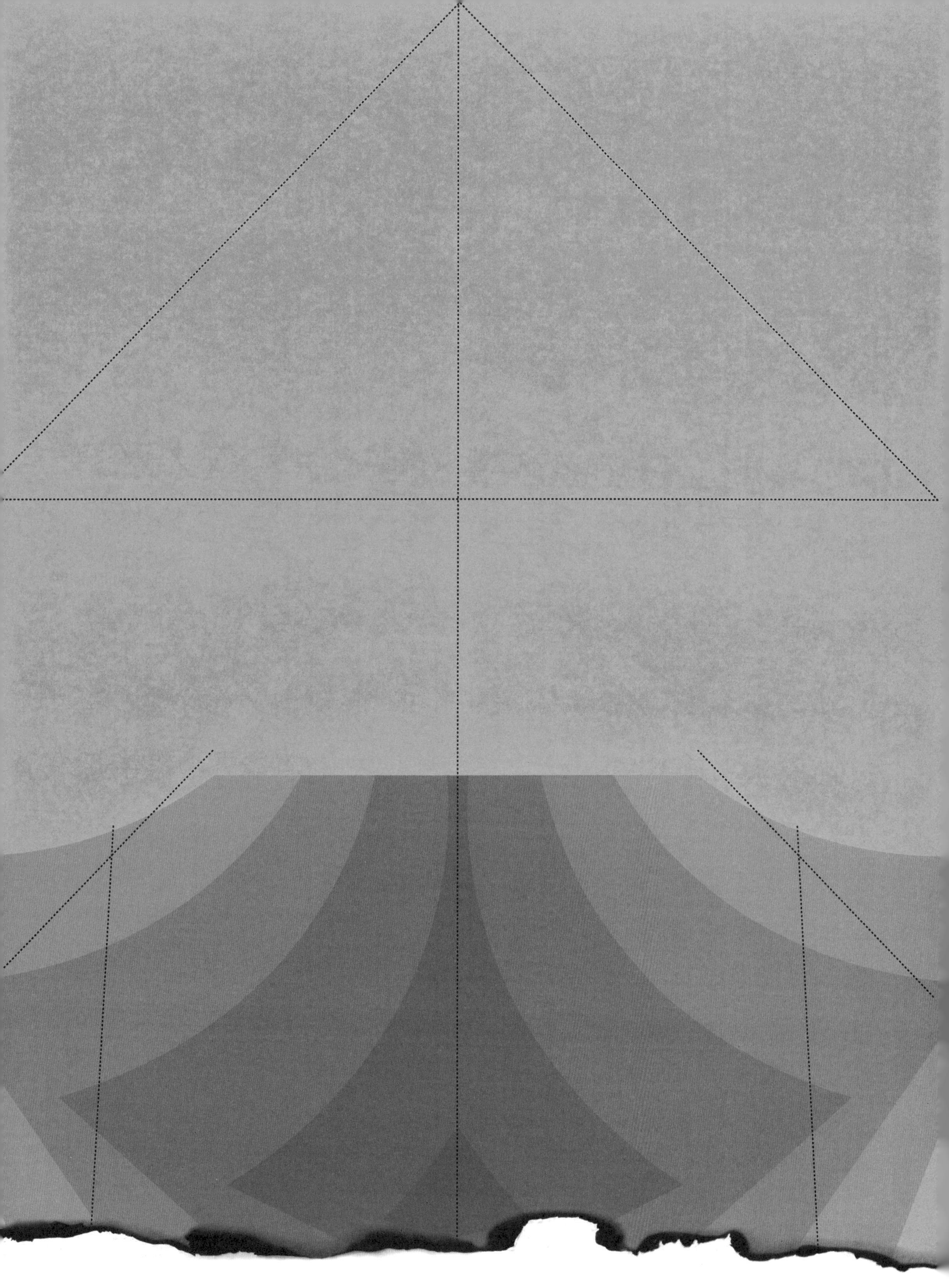

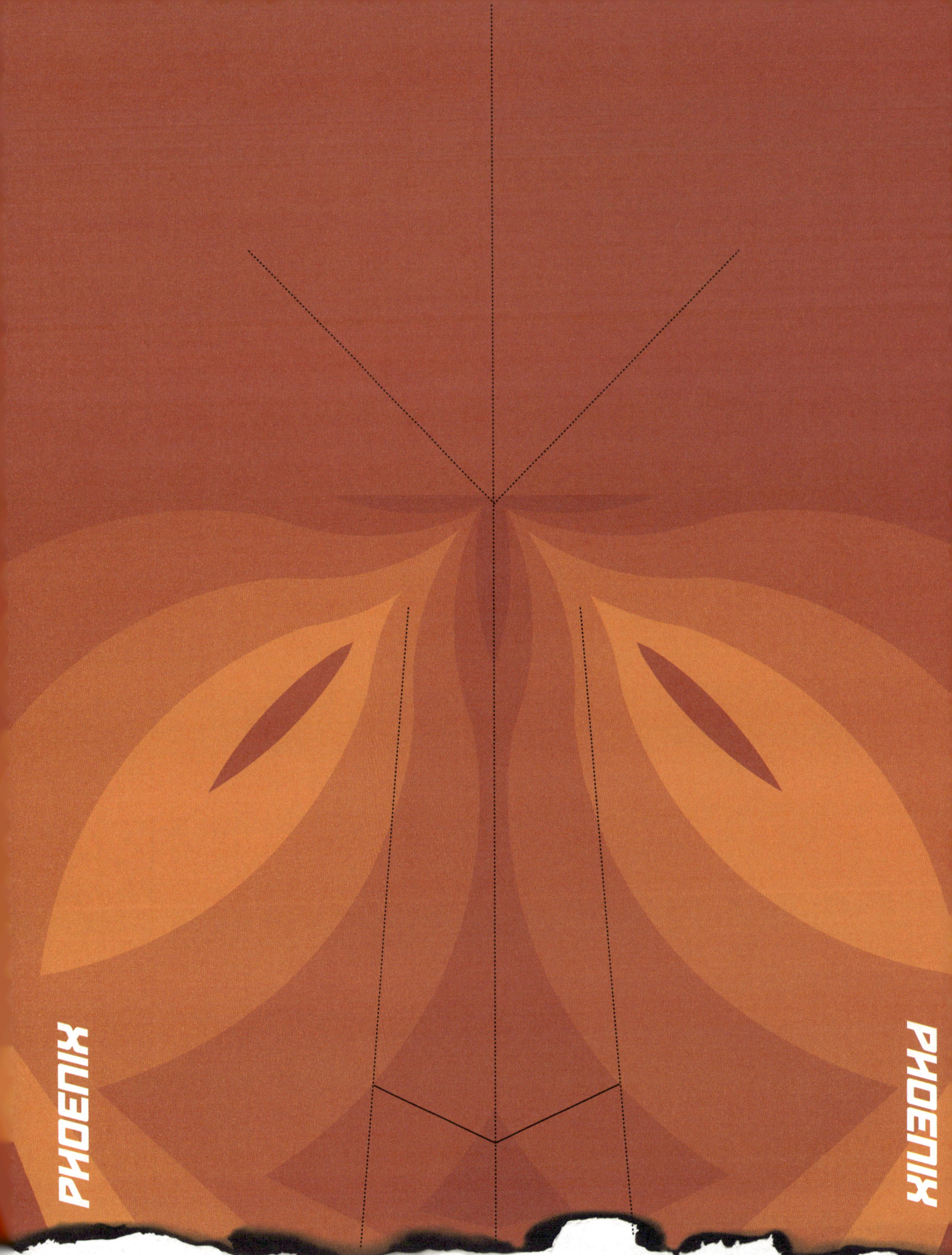
PHOENIX
PHOENIX

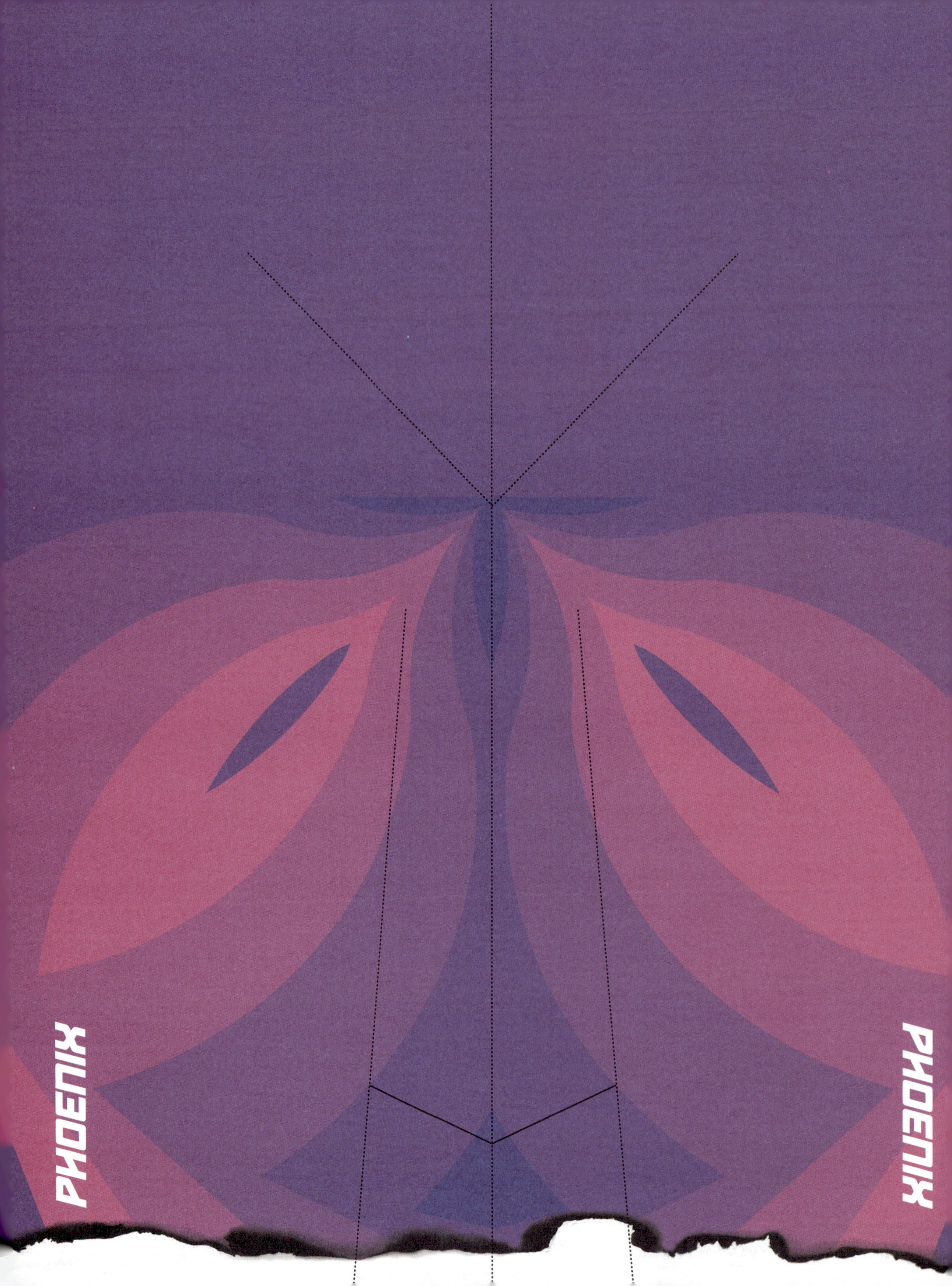
PHOENIX
PHOENIX

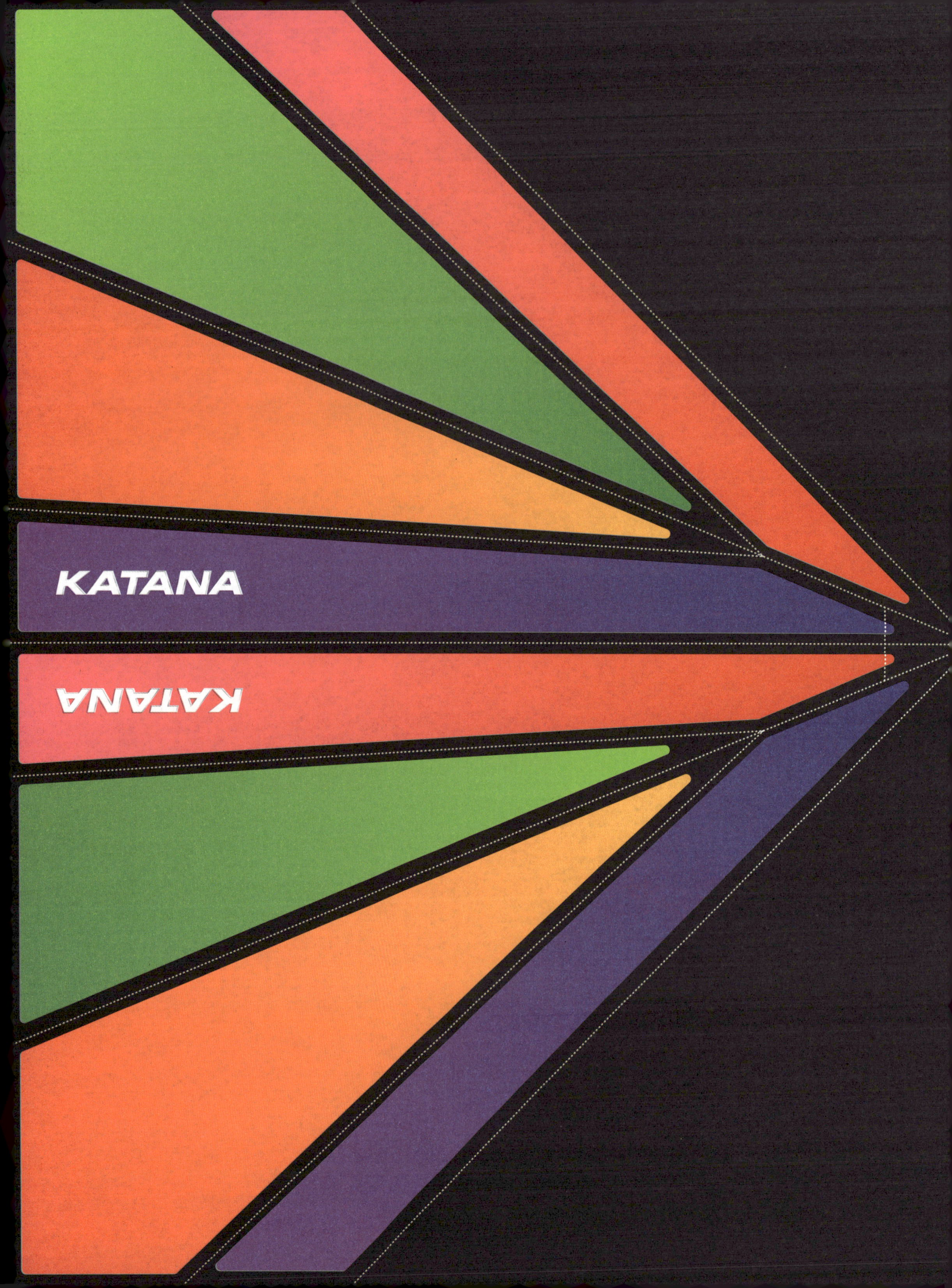
KATANA
KATANA

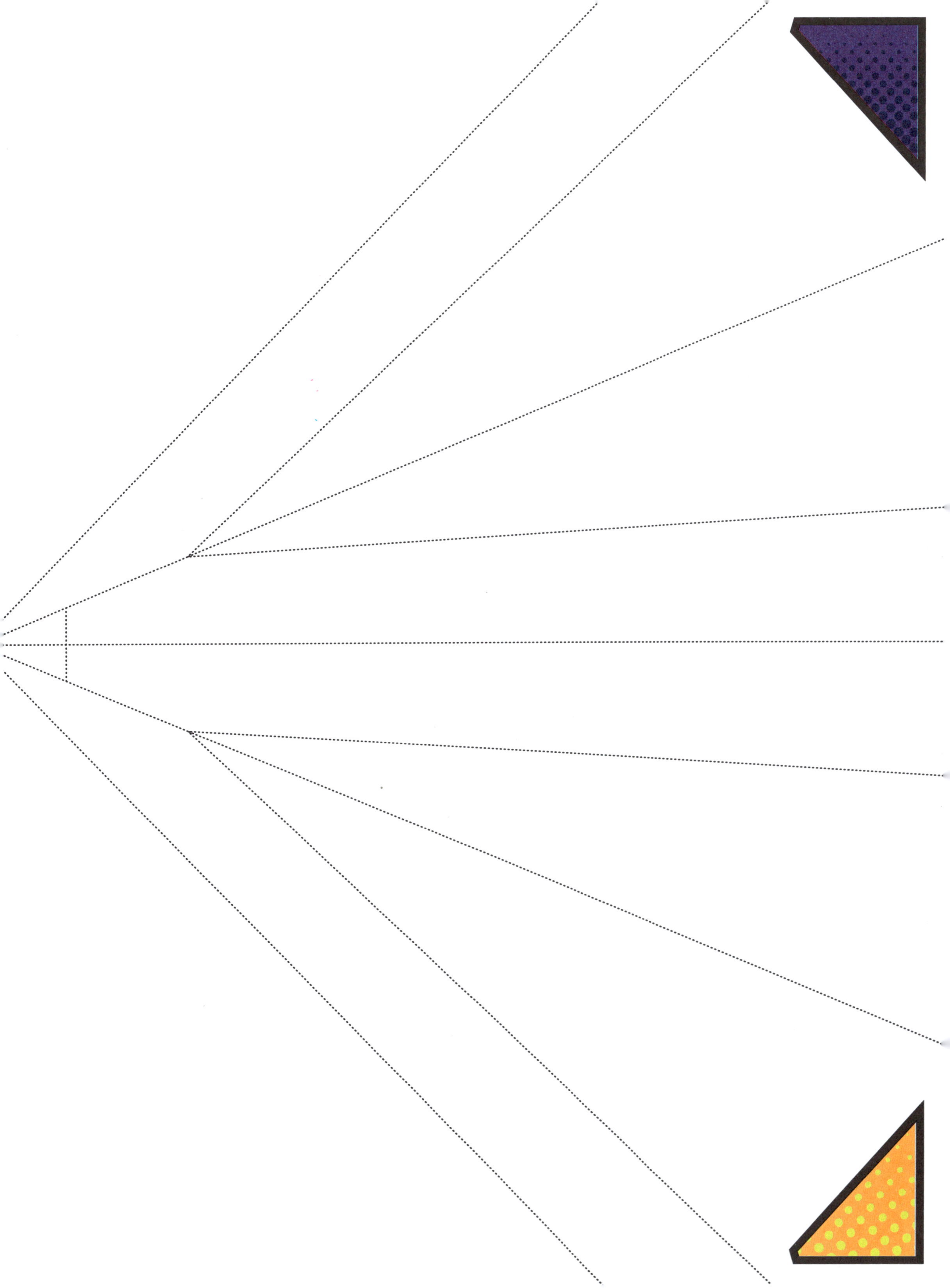

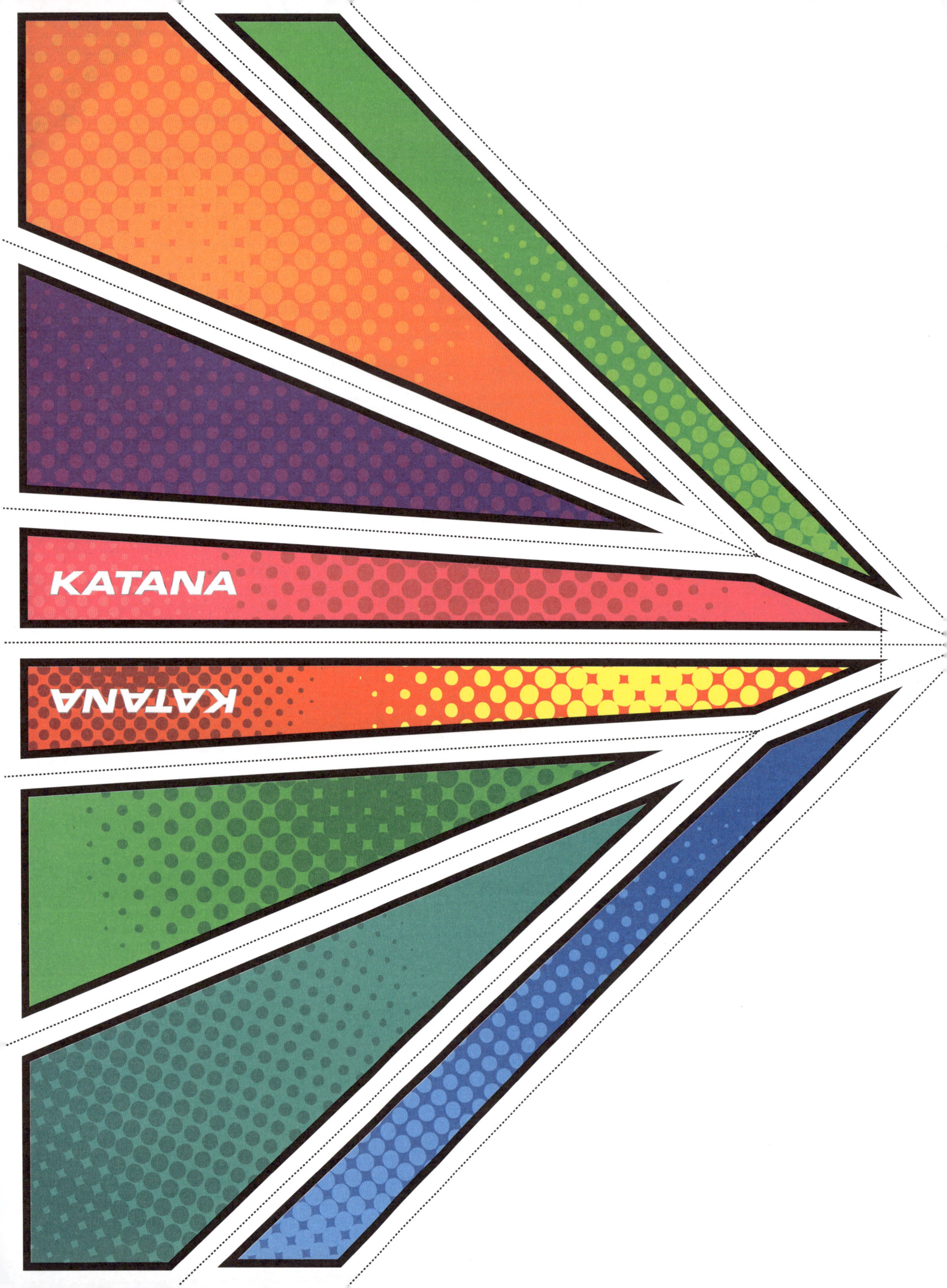
KATANA
KATANA

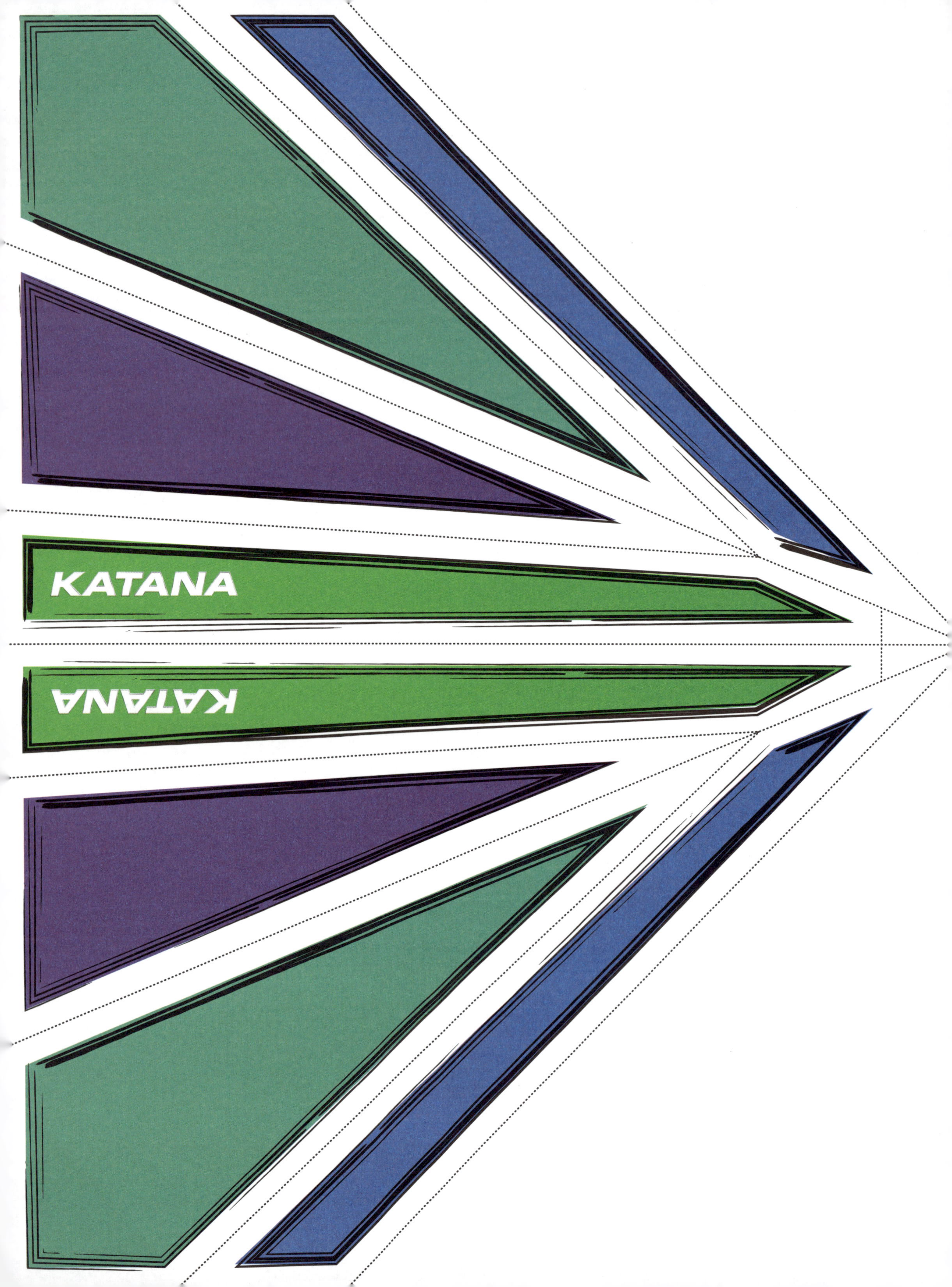
KATANA
KATANA

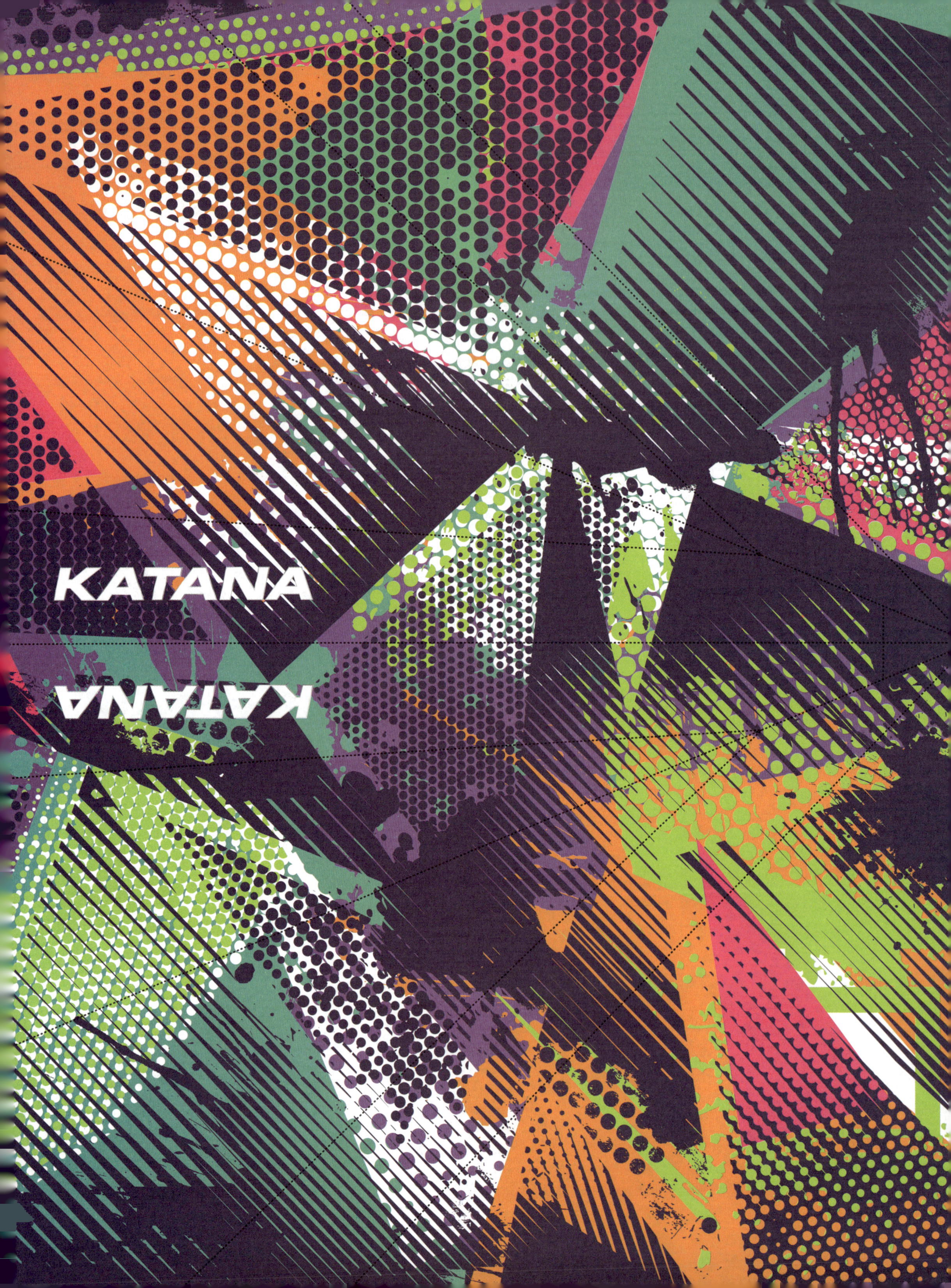
KATANA
KATANA

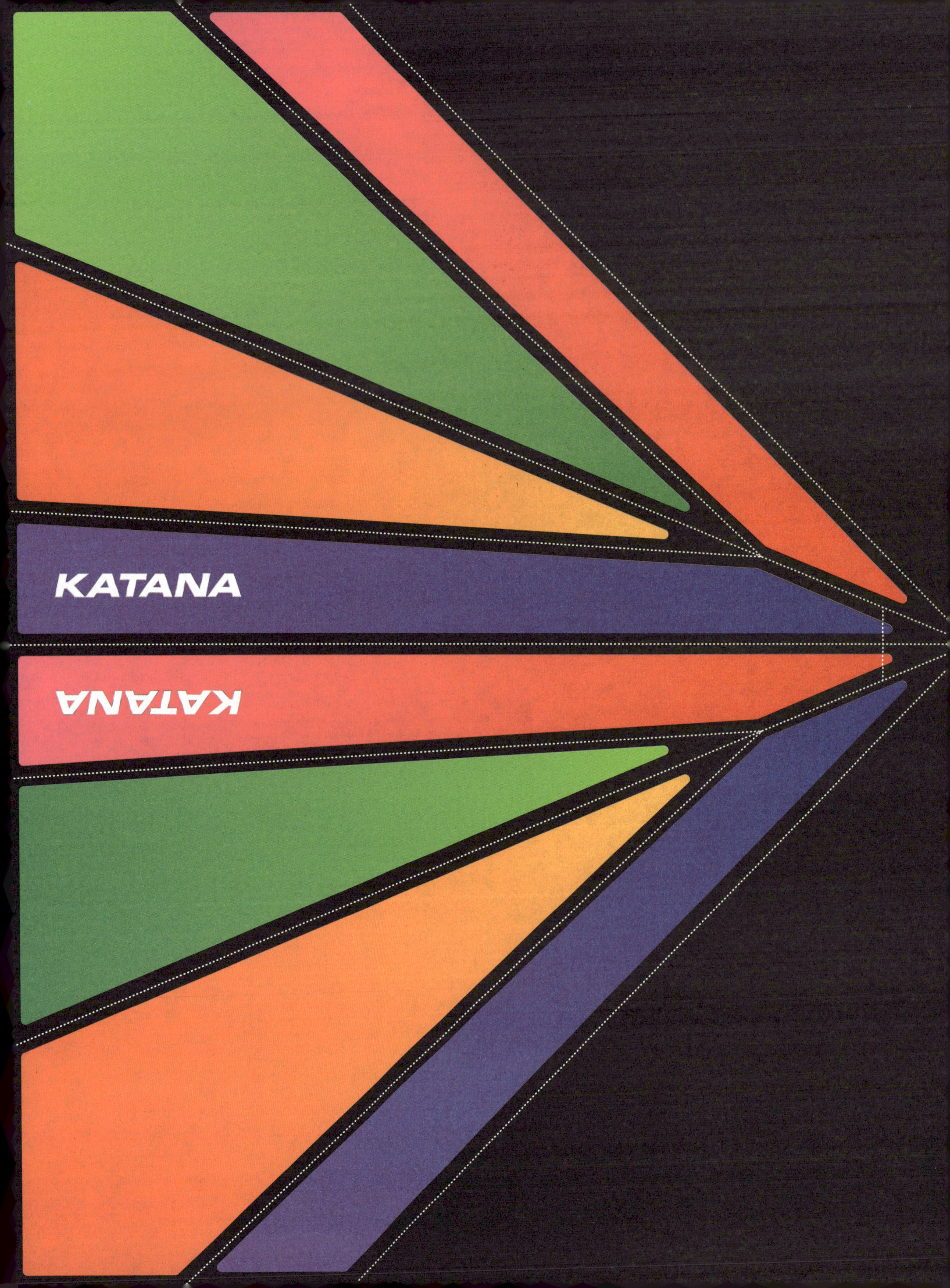
KATANA
KATANA

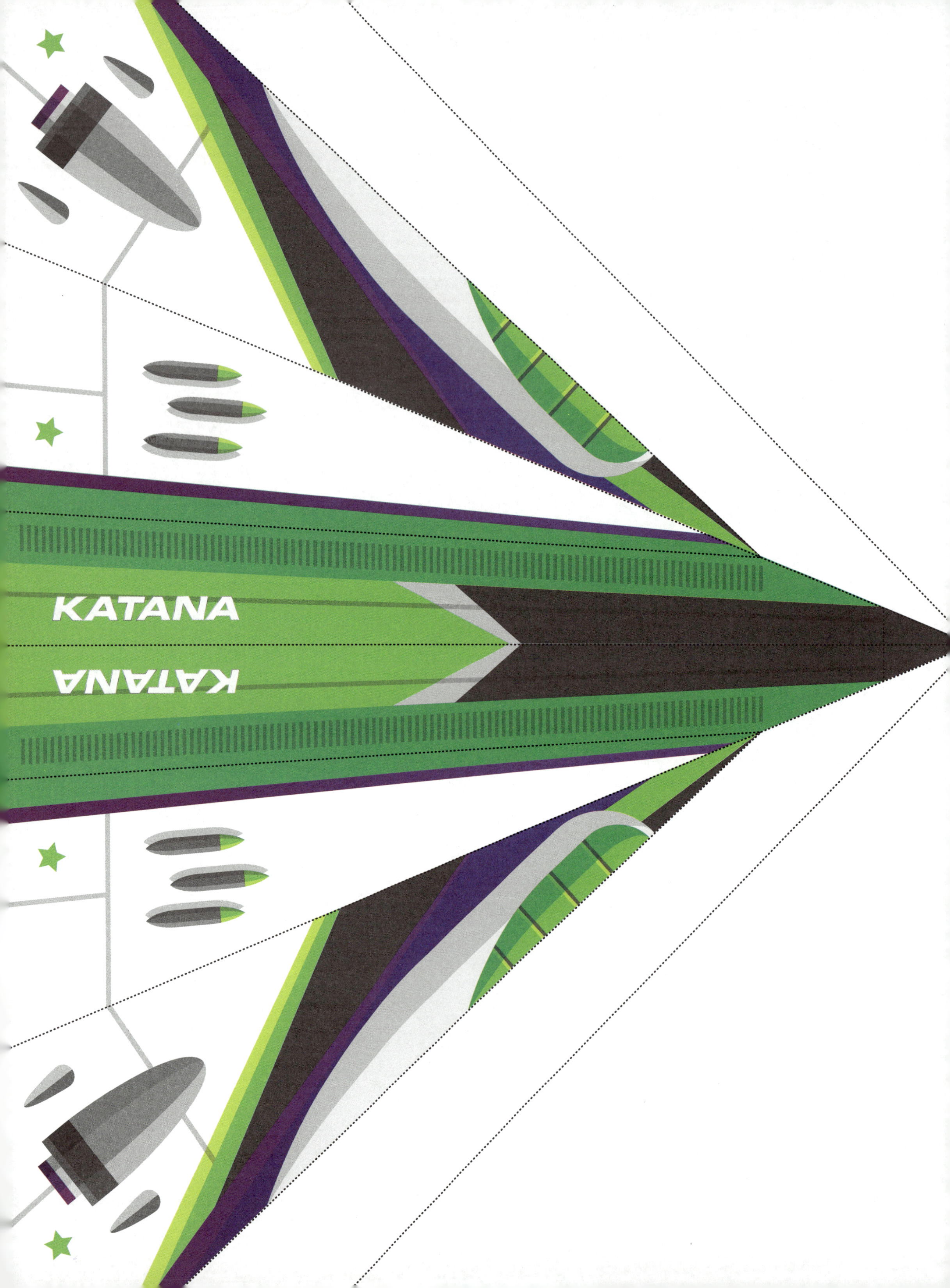
KATANA
KATANA

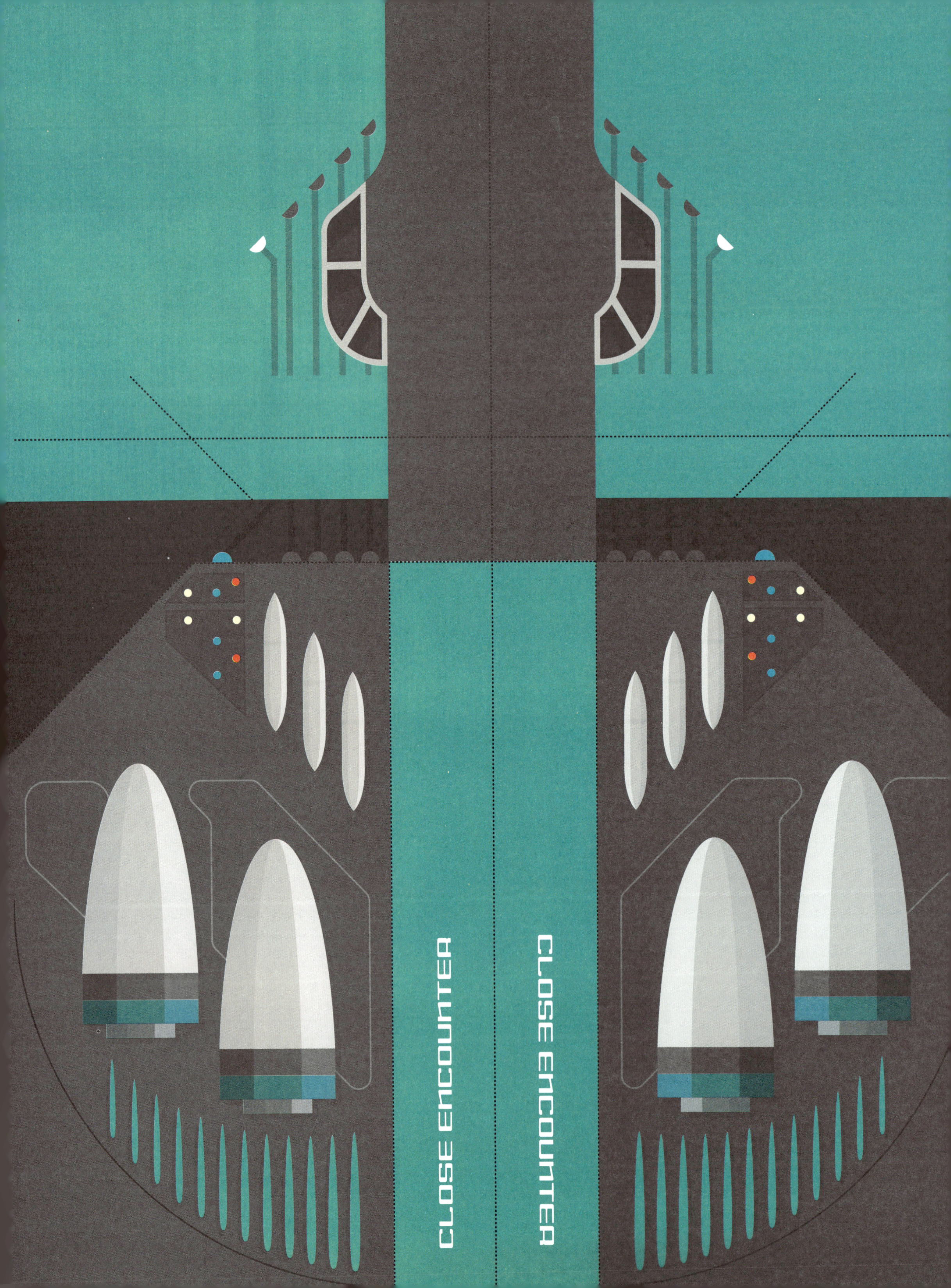
CLOSE ENCOUNTER
CLOSE ENCOUNTER

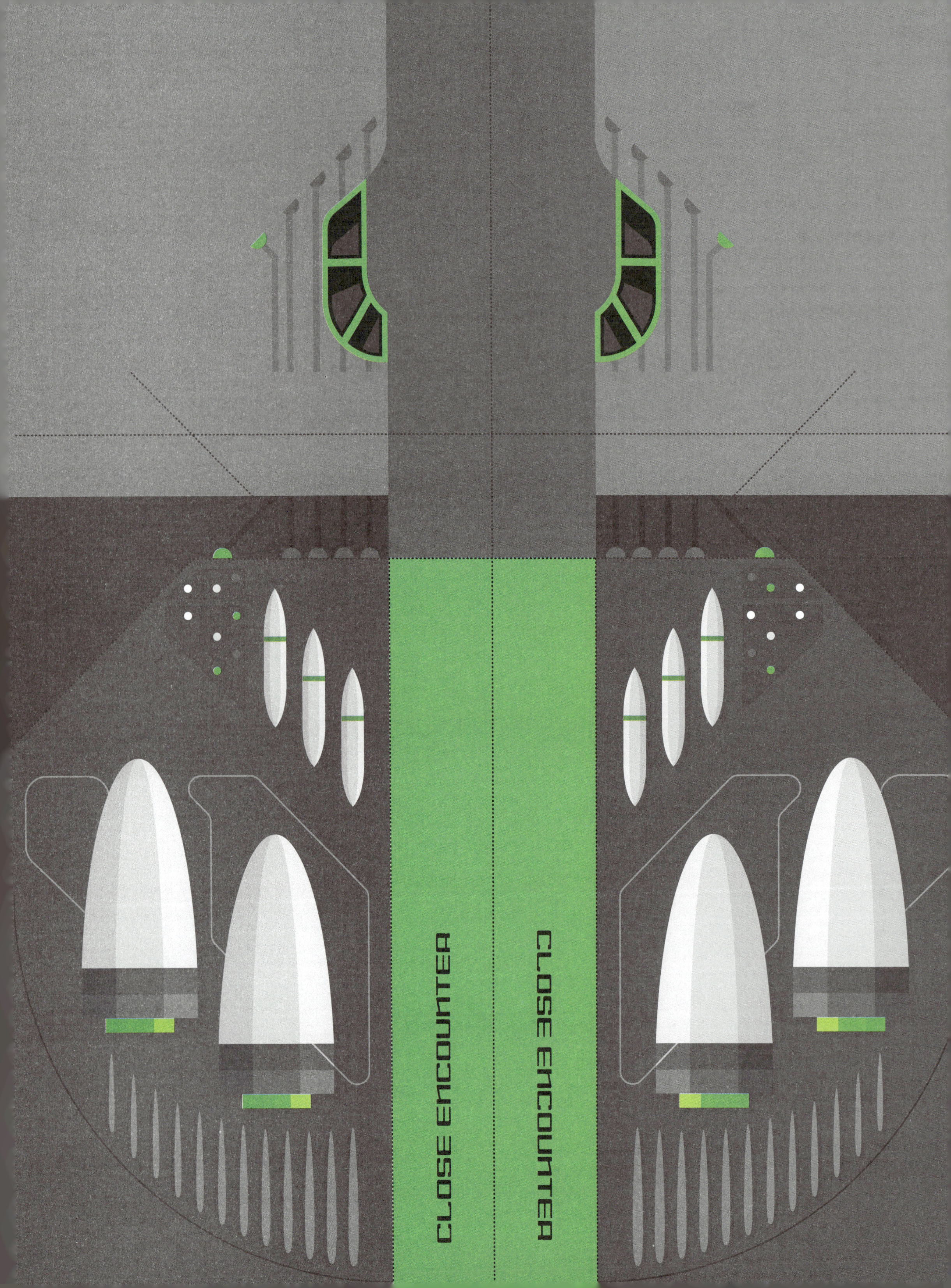
CLOSE ENCOUNTER
CLOSE ENCOUNTER

CLOSE ENCOUNTER

CLOSE ENCOUNTER

CLOSE ENCOUNTER
CLOSE ENCOUNTER

CLOSE ENCOUNTER
CLOSE ENCOUNTER

TANGENT

TANGENT

TANGENT
TANGENT

TANGENT

TANGENT

United States
United States
Orbiter

Orbiter

Orbiter

Orbiter

Orbiter

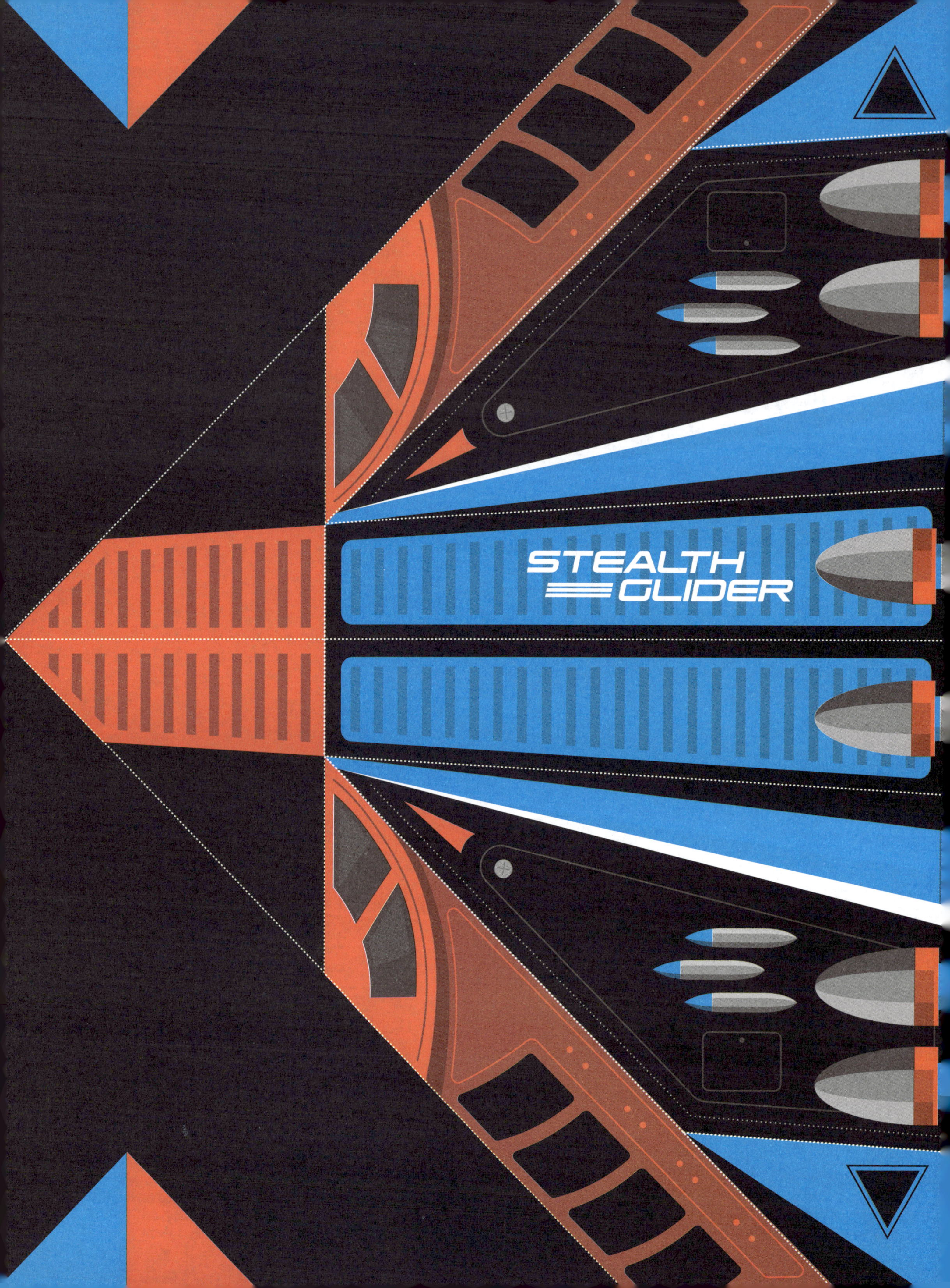
STEALTH
GLIDER

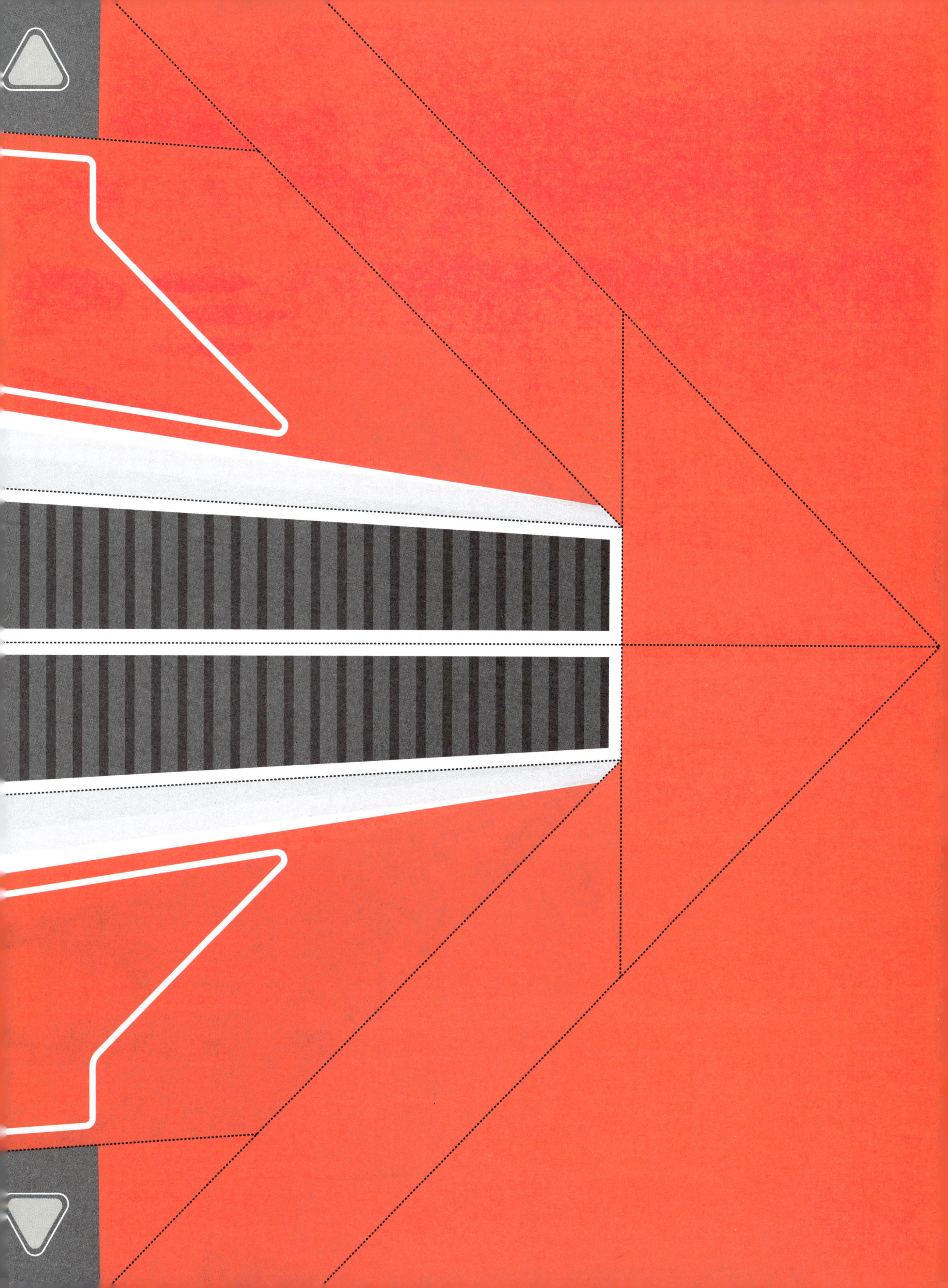

STEALTH
GLIDER

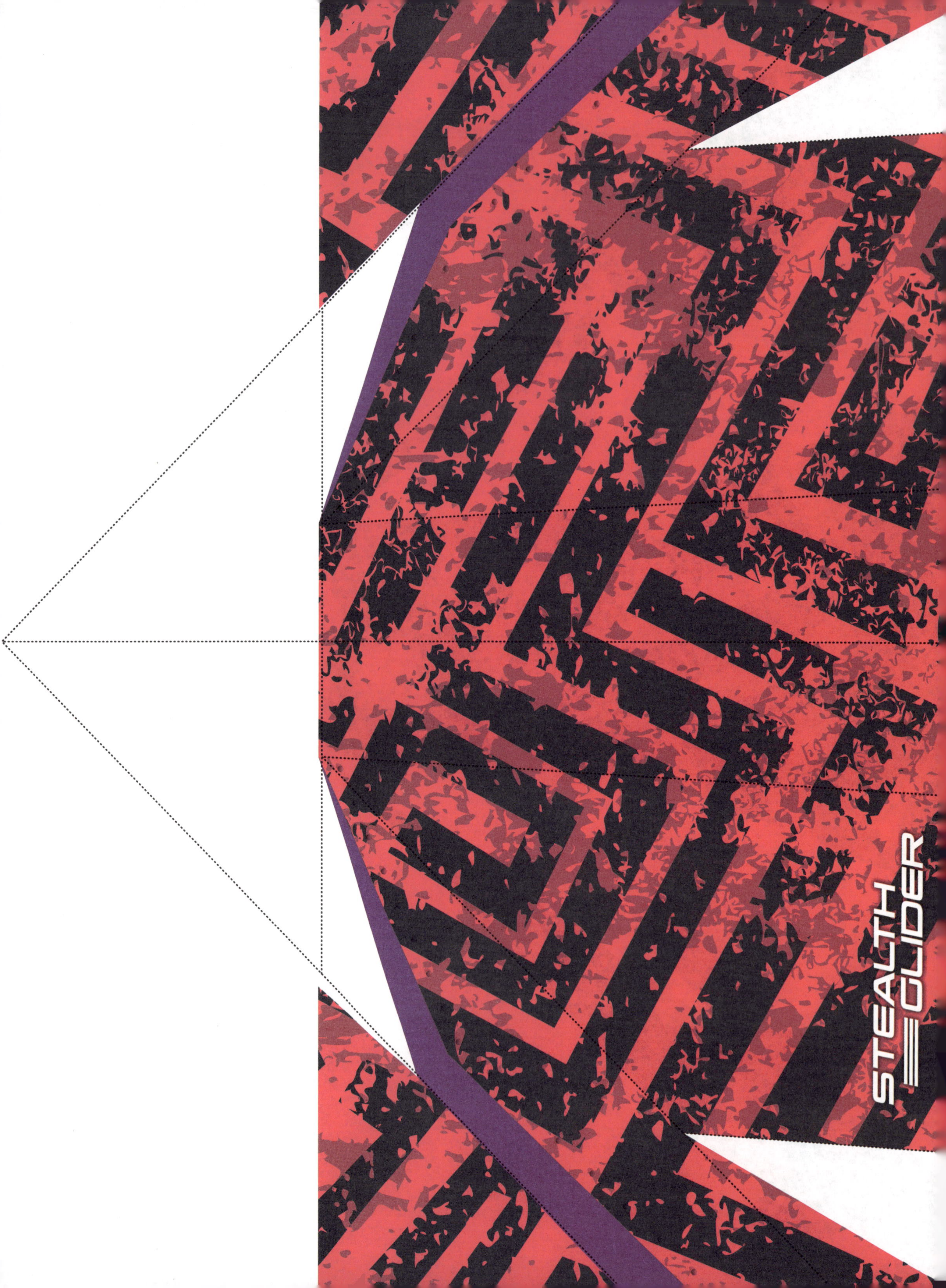
STEALTH
GLIDER

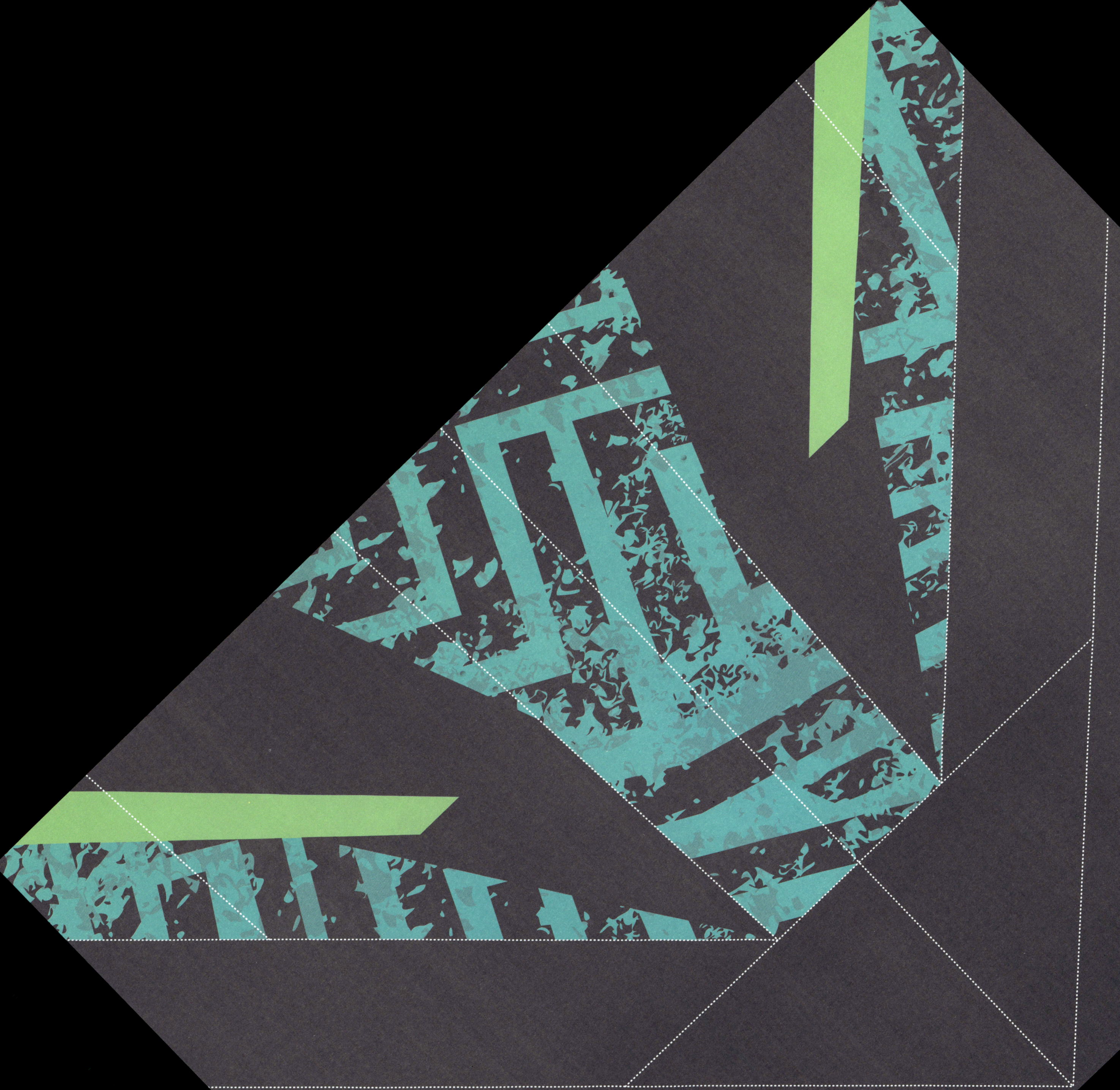

STEALTH GLIDER

STEALTH GLIDER

REGAL EAGLE

REGAL EAGLE

REGAL EAGLE

REGAL EAGLE

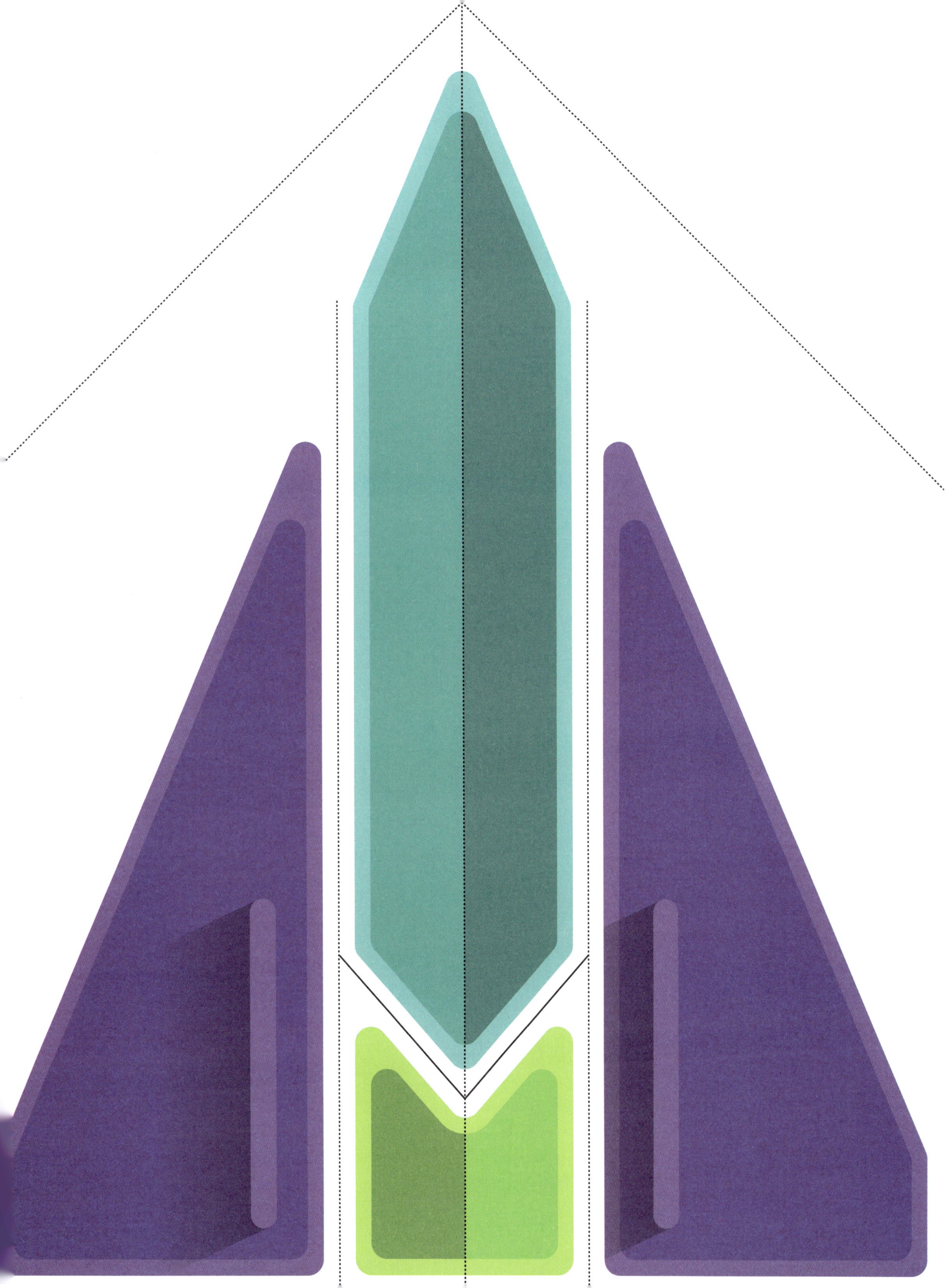

REGAL
EAGLE

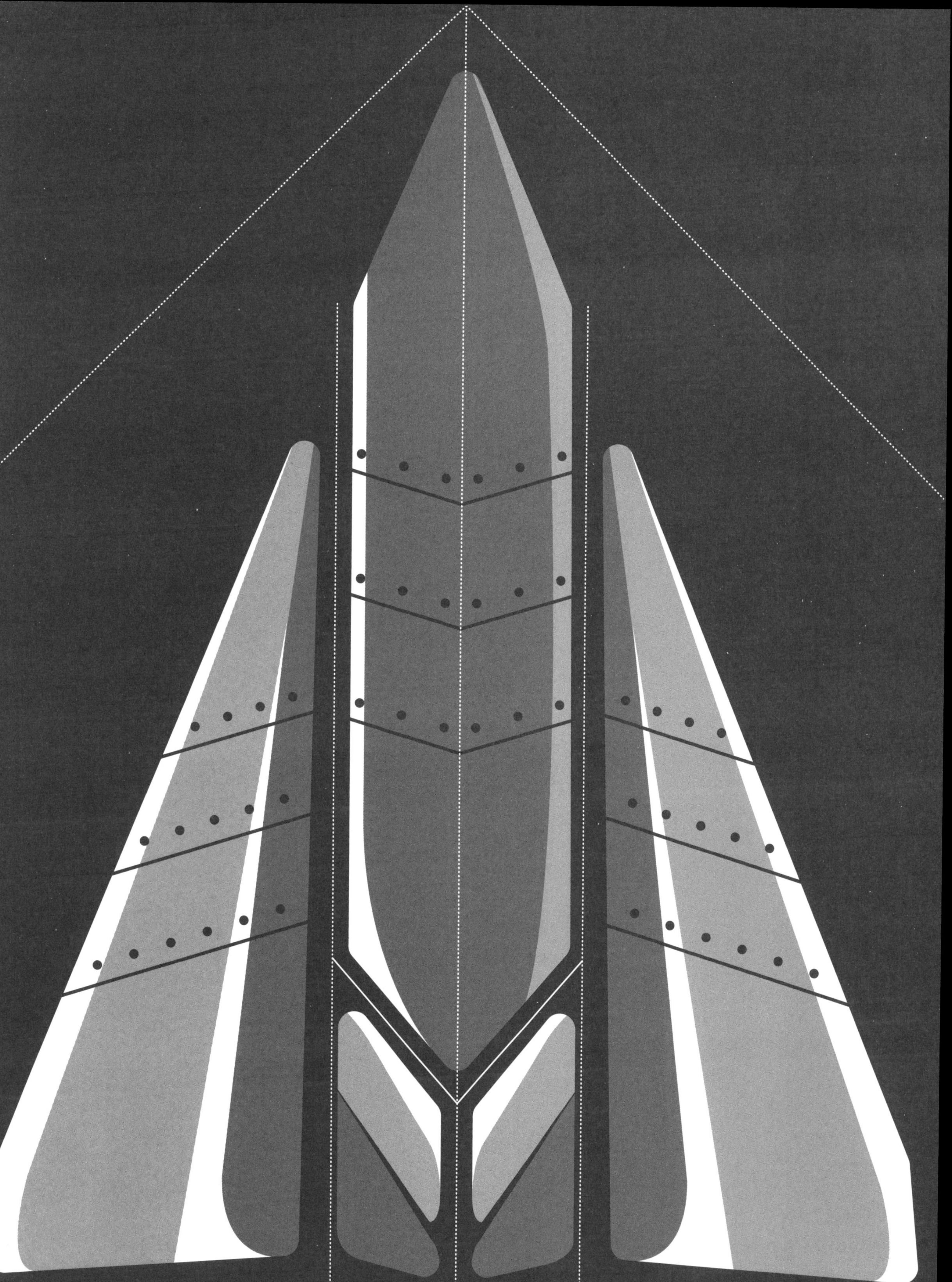

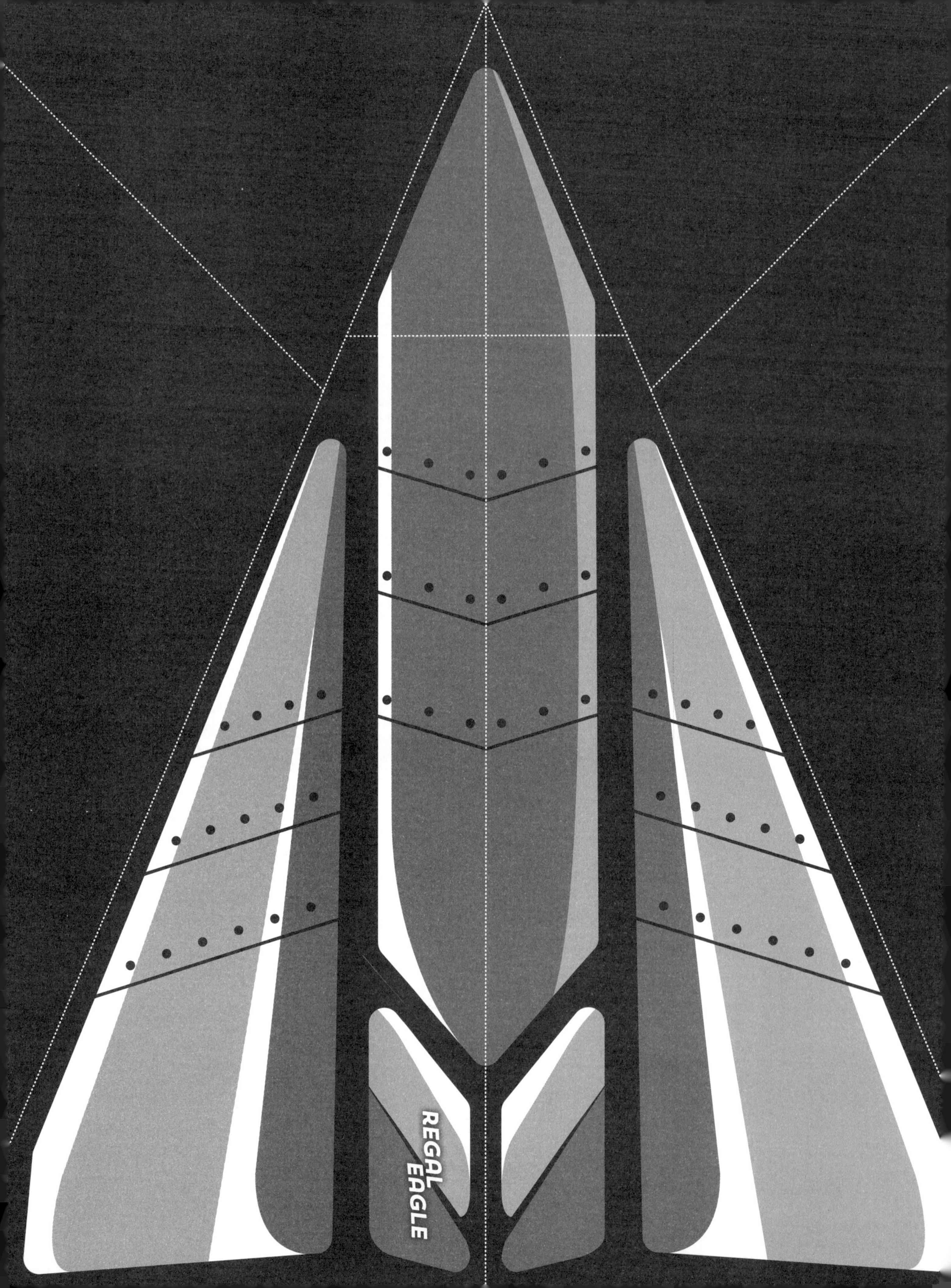
REGAL
EAGLE

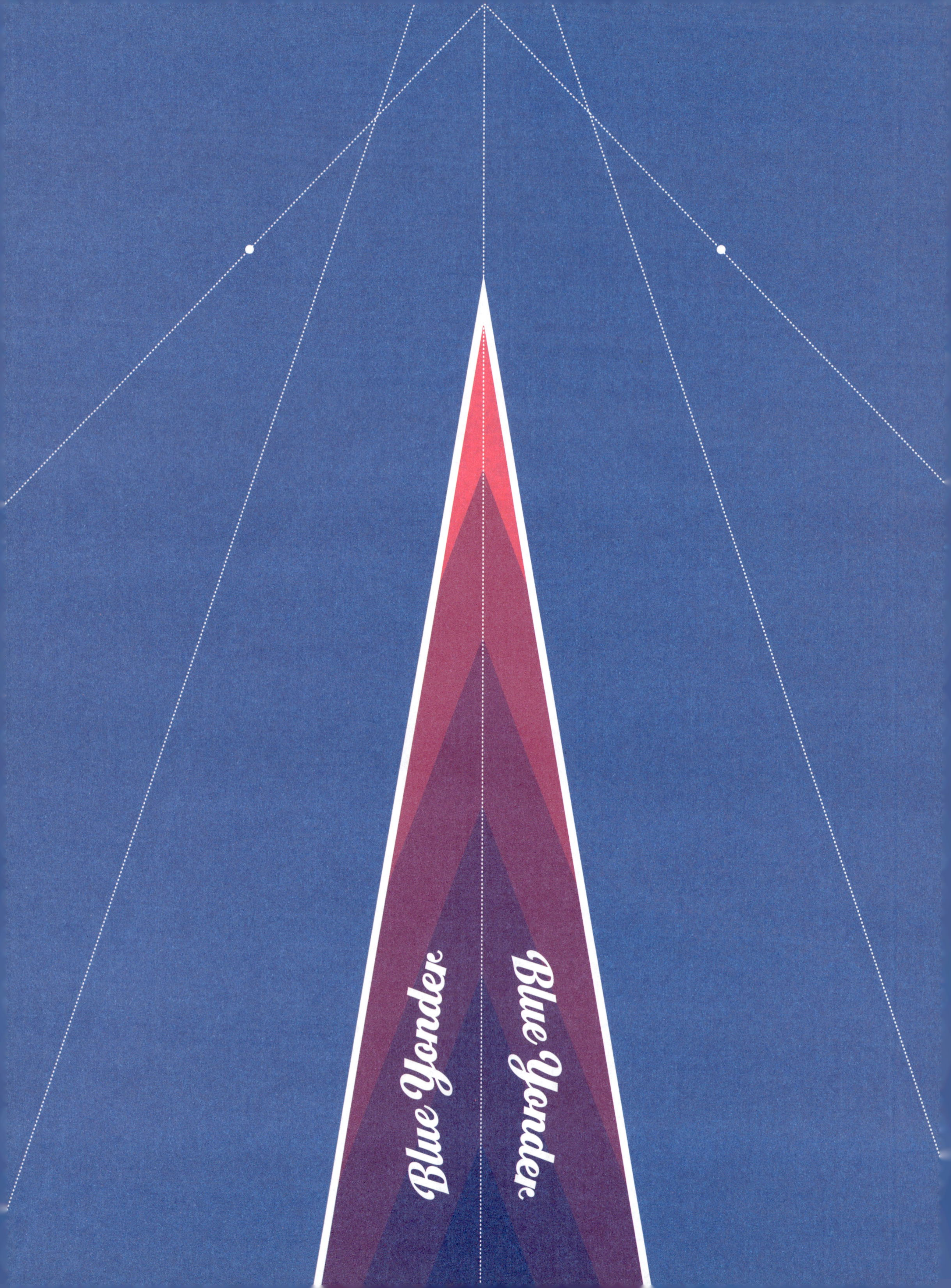
Blue Yonder
Blue Yonder

Blue Yonder
Blue Yonder

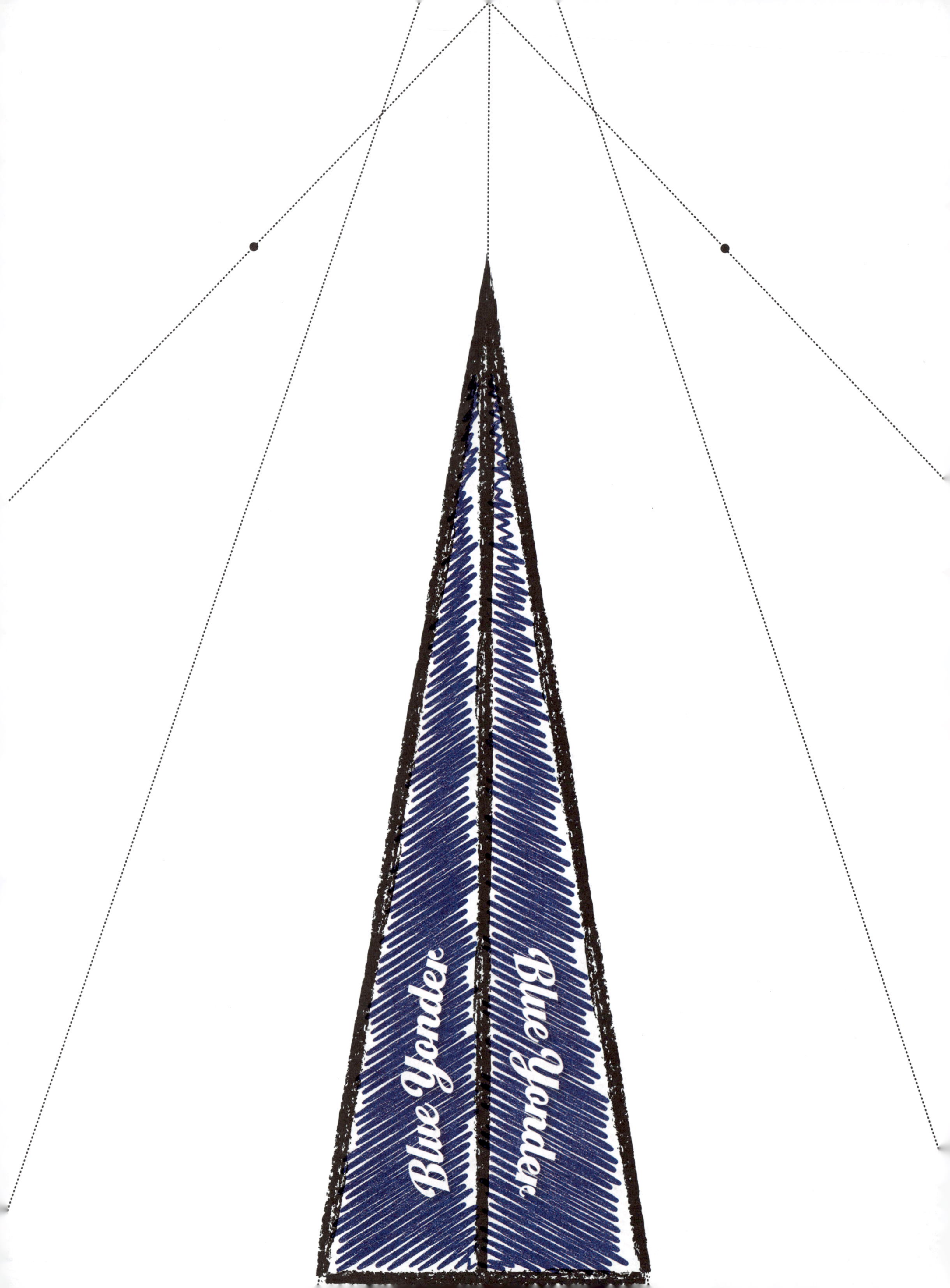
Blue Yonder
Blue Yonder

Blue Yonder
Blue Yonder

Blue Yonder

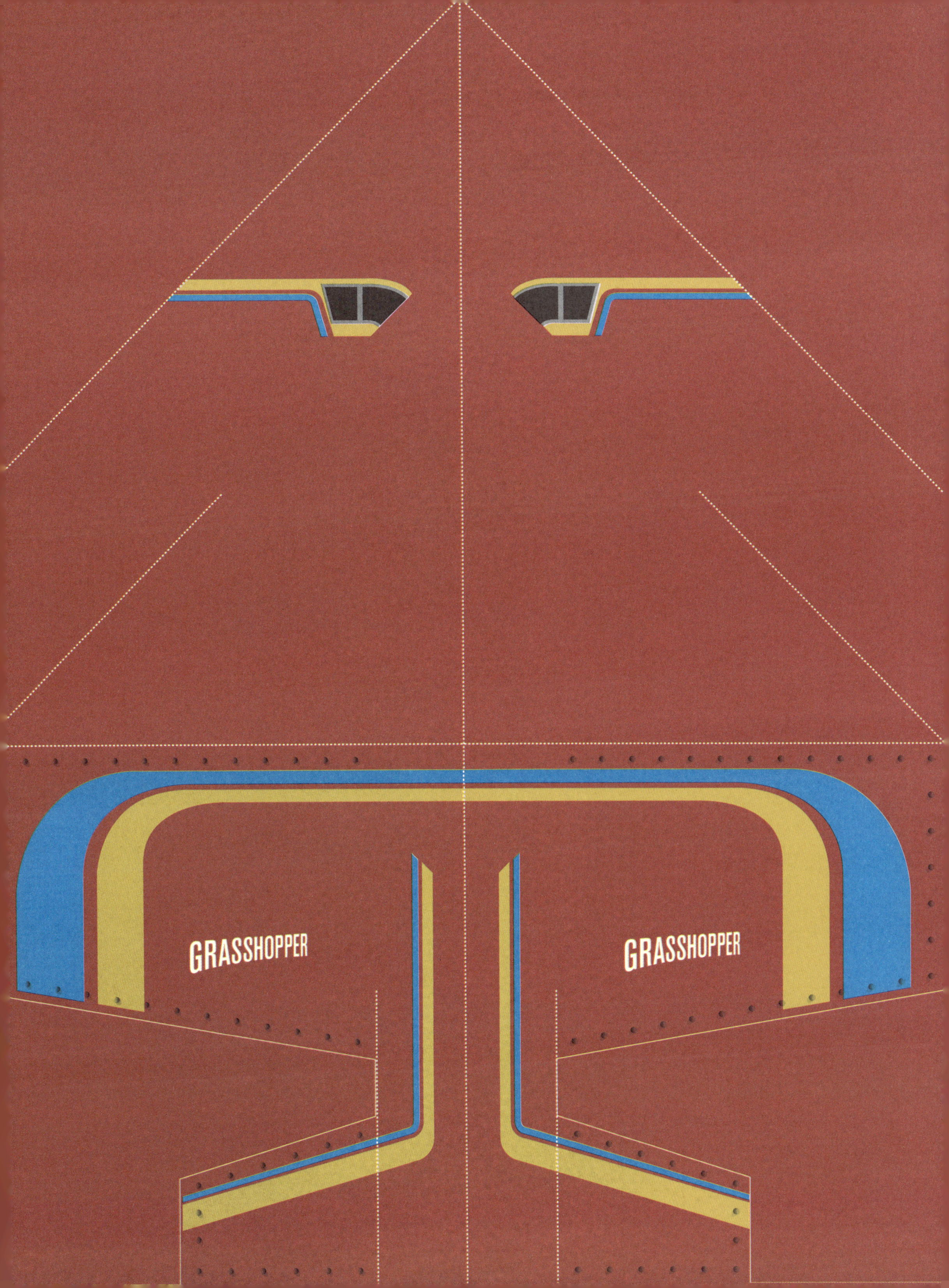
GRASSHOPPER
GRASSHOPPER

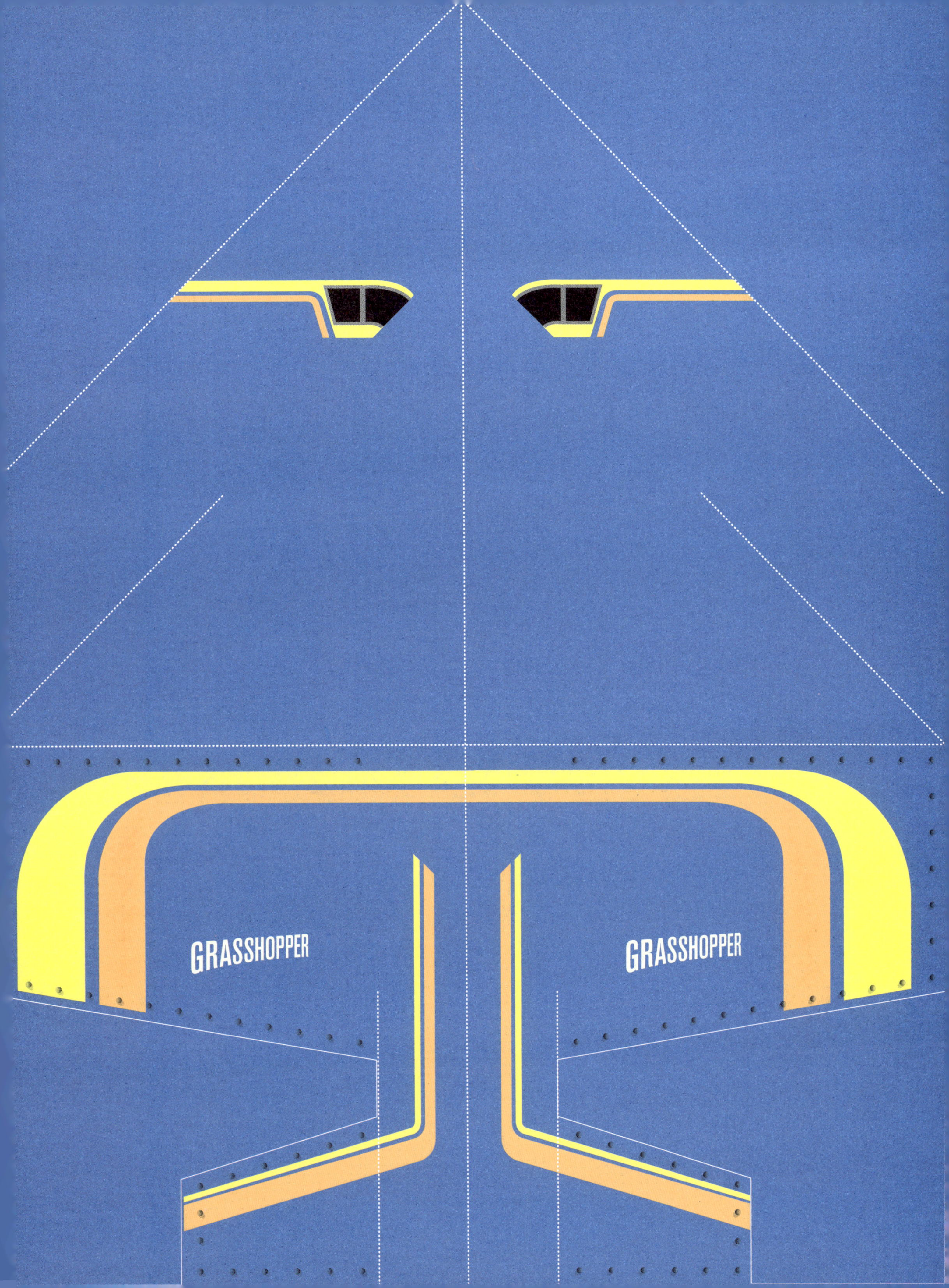
GRASSHOPPER
GRASSHOPPER

GRASSHOPPER
GRASSHOPPER

GRASSHOPPER
GRASSHOPPER

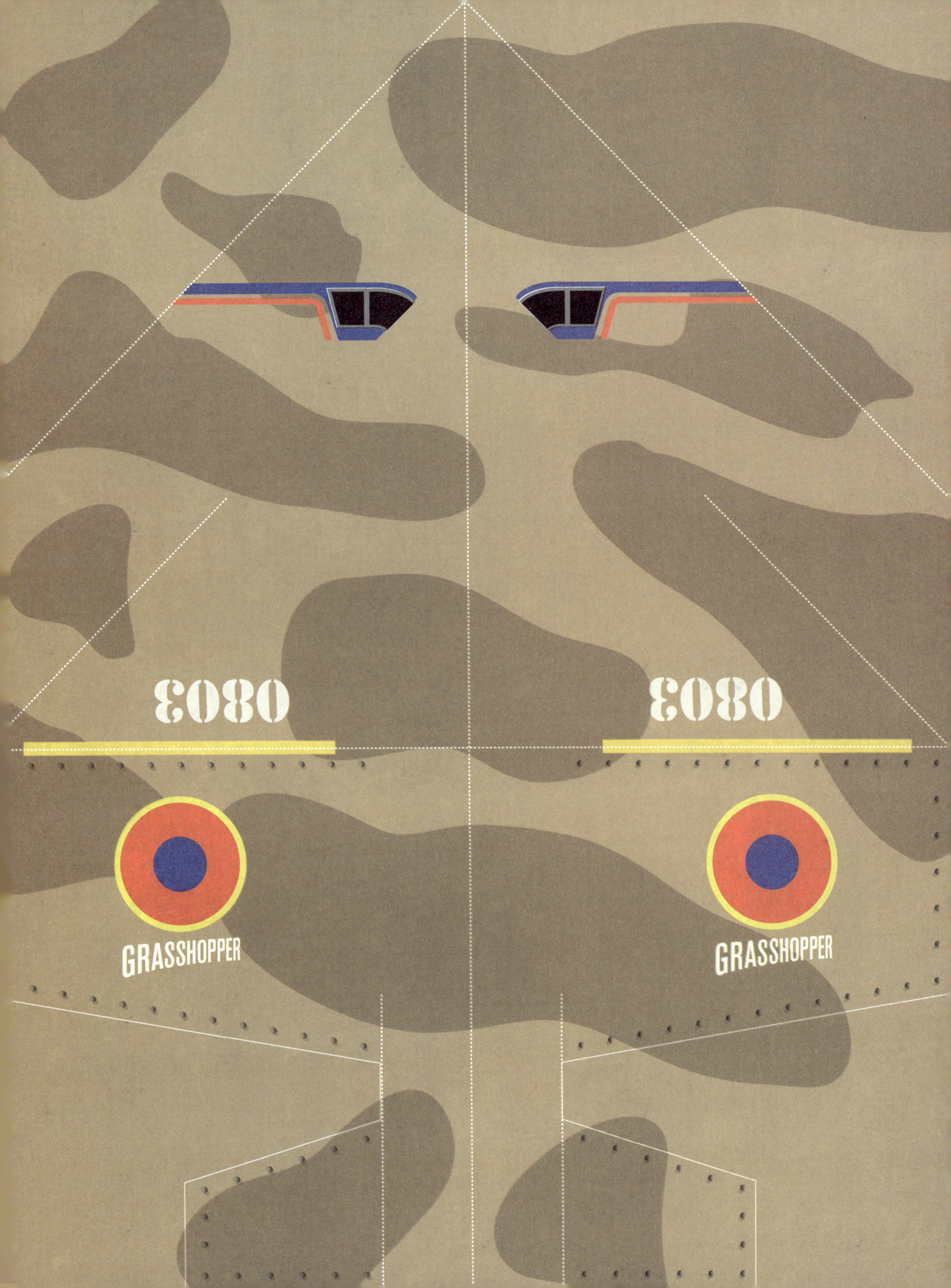
GRASSHOPPER
GRASSHOPPER

FLEUR DE LOOP

FLEUR DE LOOP

FLEUR DE LOOP

FLEUR DE LOOP

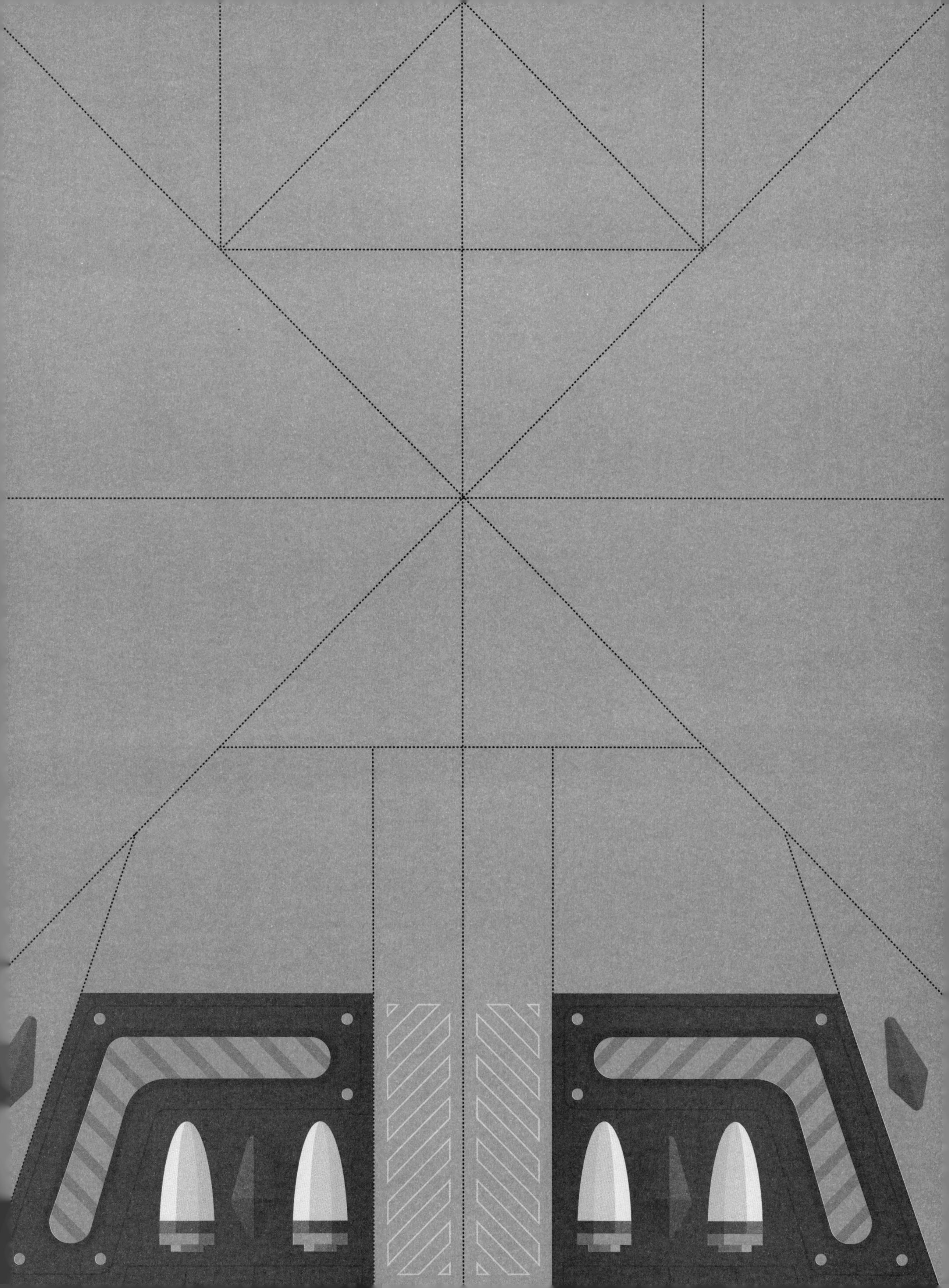

FLEUR DE LOOP

FLEUR DE LOOP

Jolly Dodger

Jolly Dodger

Jolly Dodger

Jolly Dodger

Jolly Dodger
Jolly Dodger

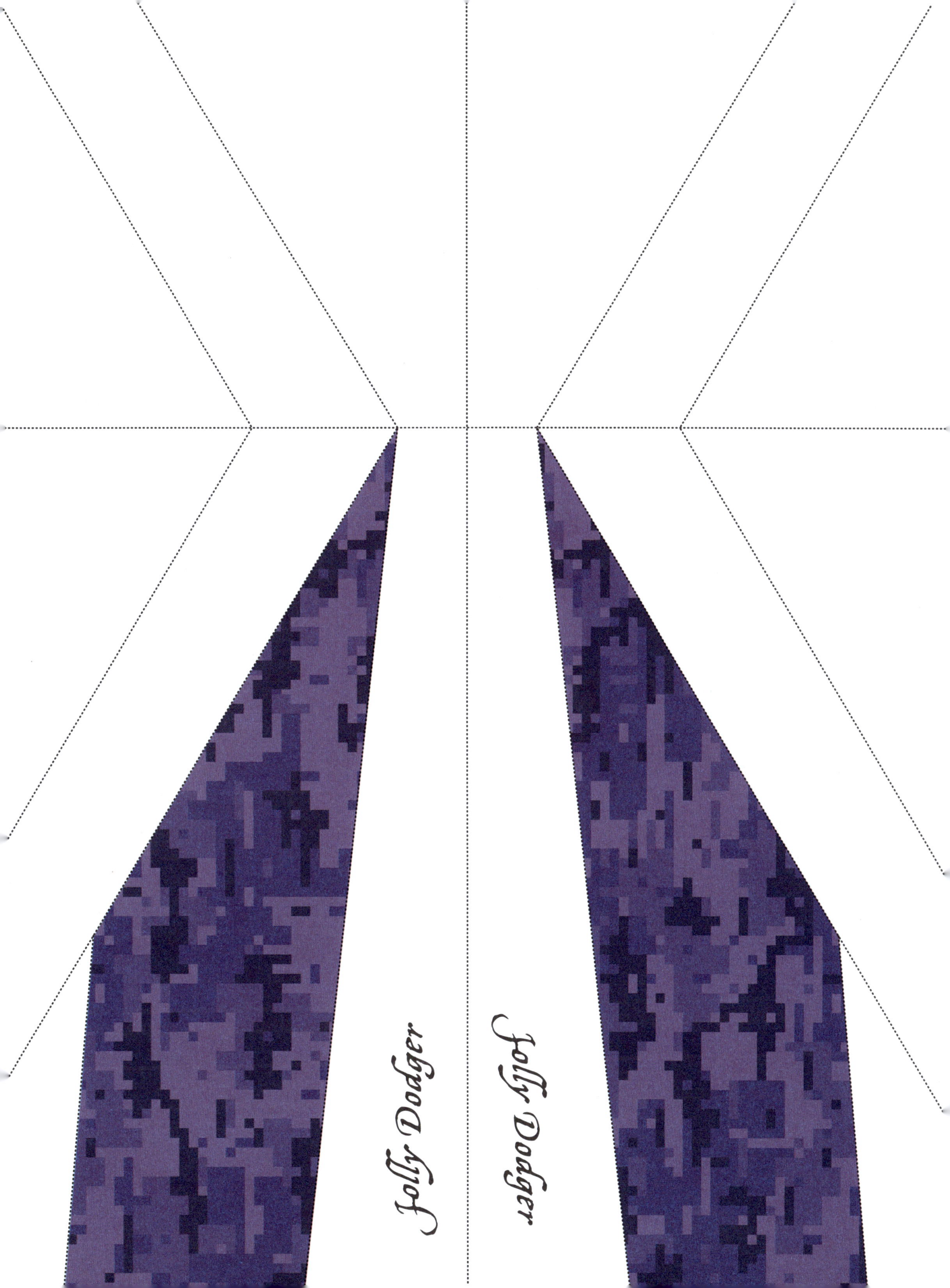
Jolly Dodger
Jolly Dodger

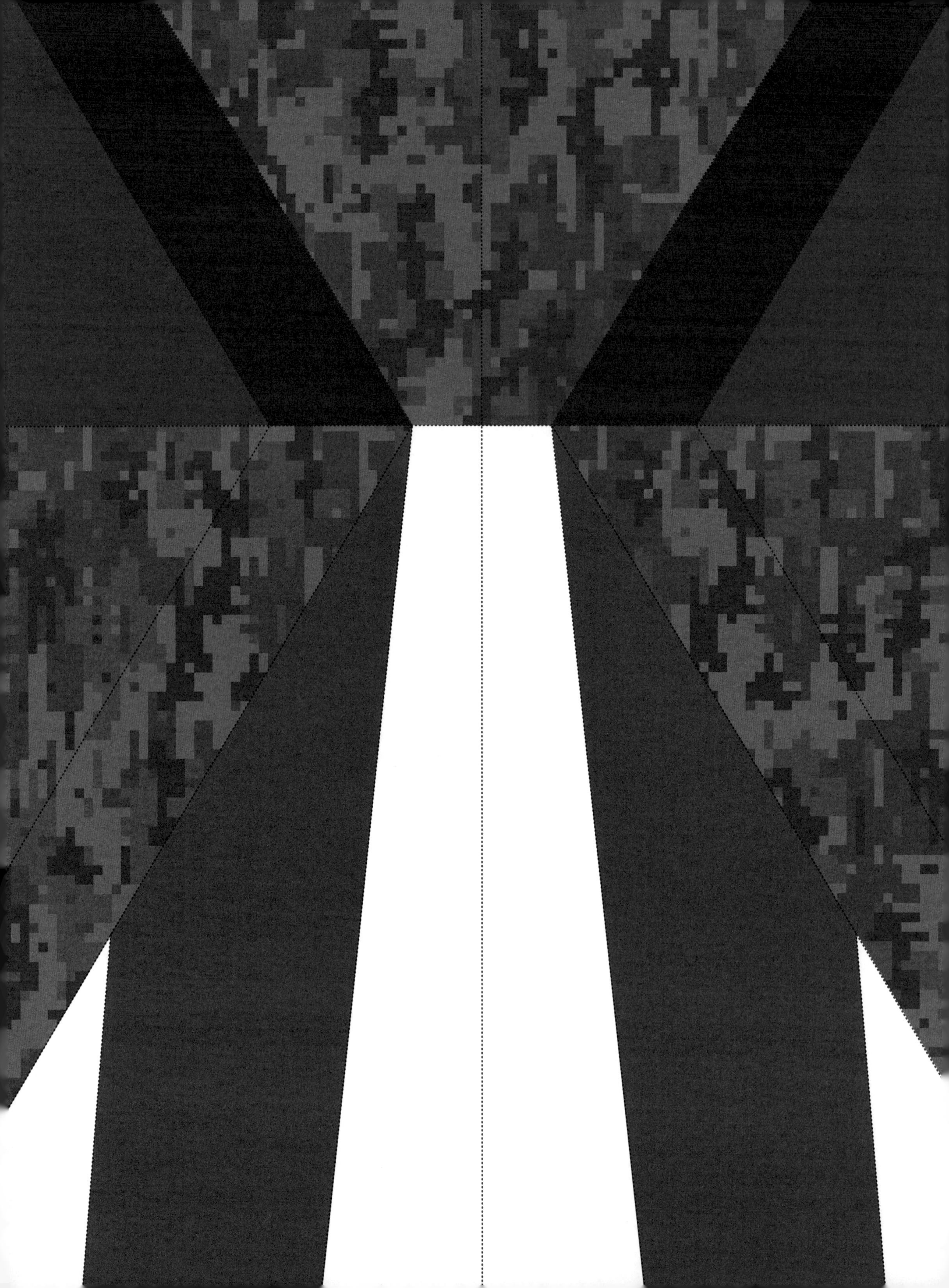

Jolly Dodger
Jolly Dodger

Jolly Dodger
Jolly Dodger

Swerving Dervish

Swerving Dervish

Swerving Dervish

Swerving Dervish

Swerving
Dervish

BIG MOUTH

BIG MOUTH

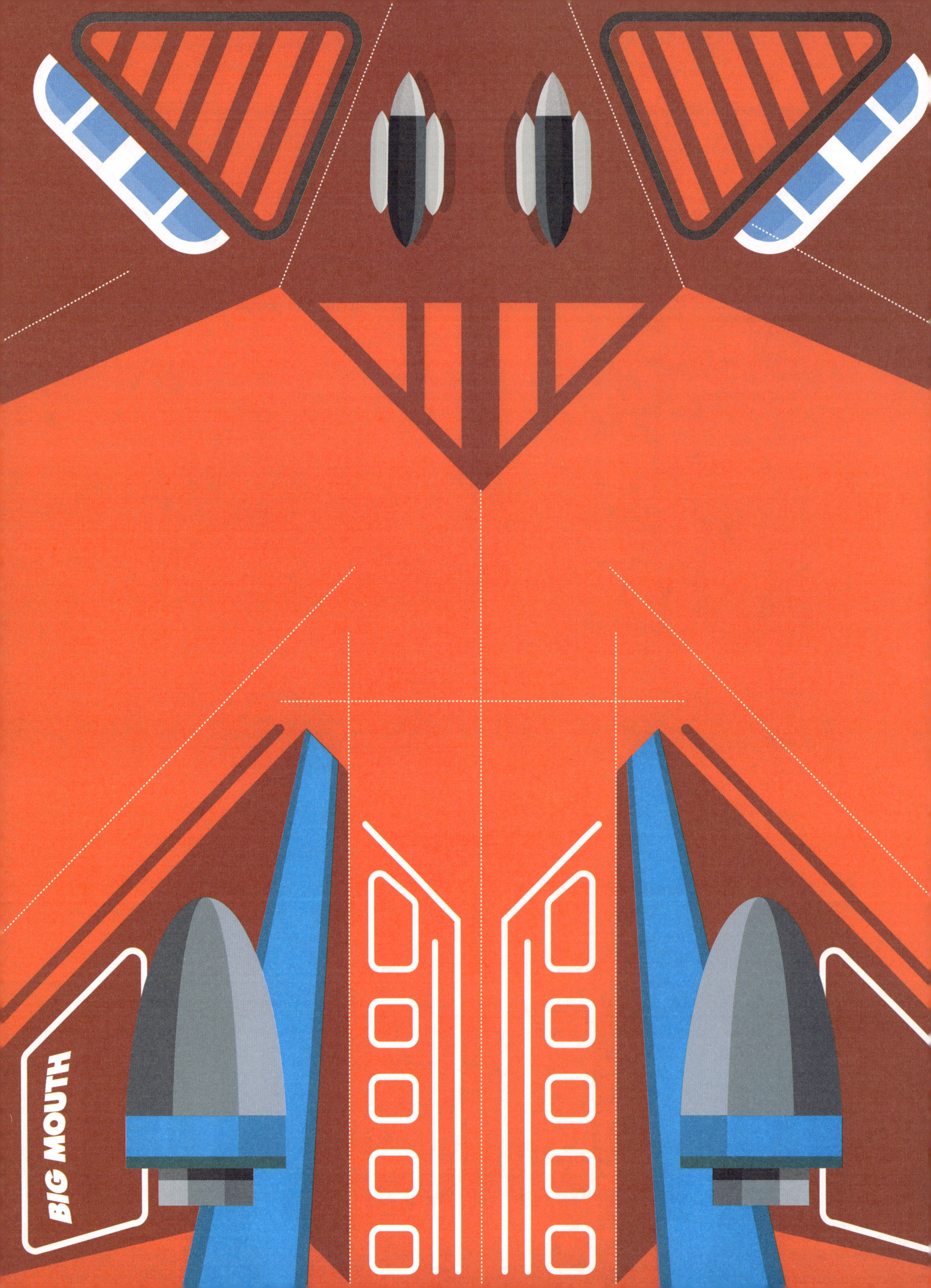
BIG MOUTH

BIG MOUTH

BIG MOUTH

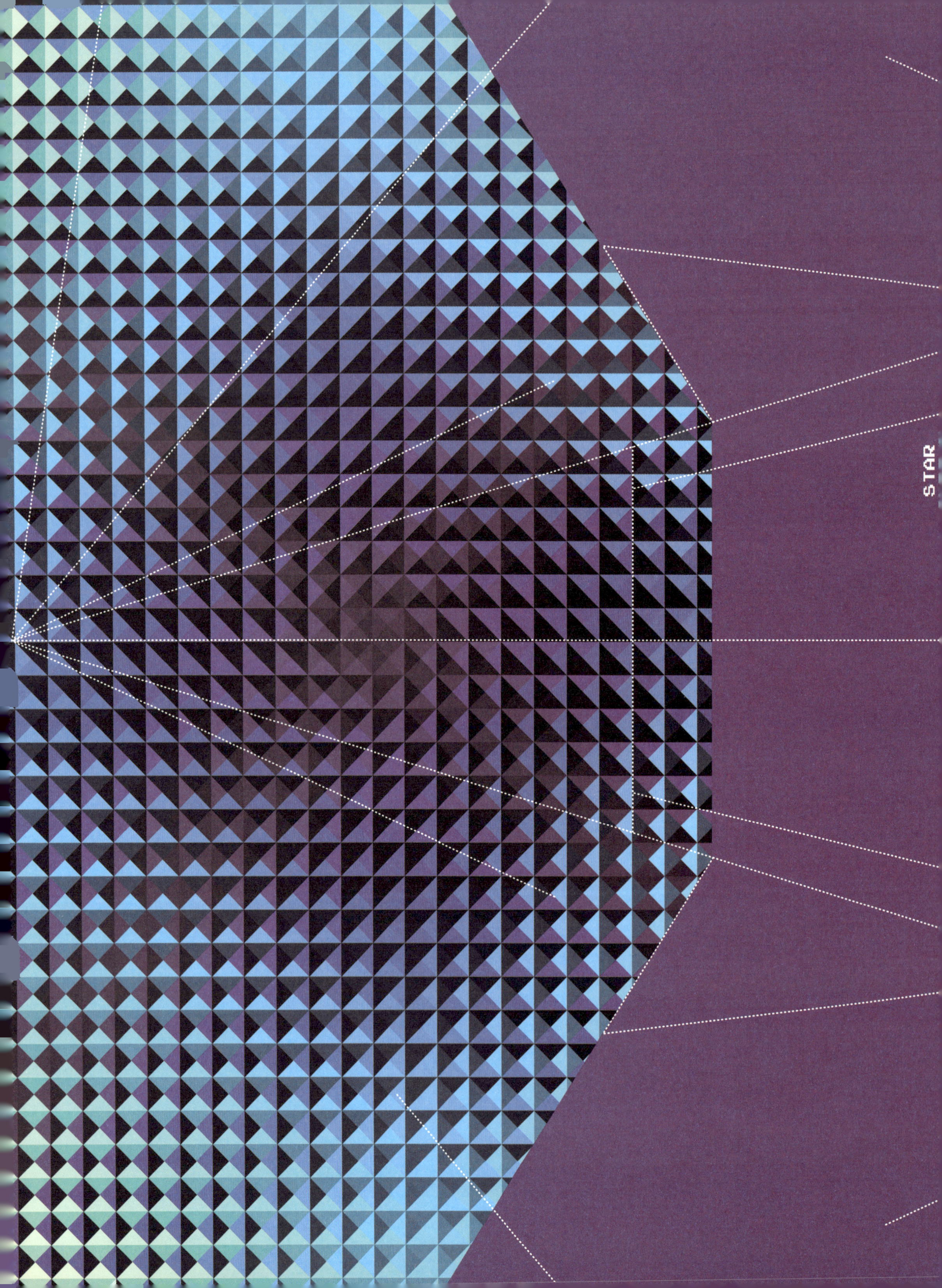
STAR

STAR

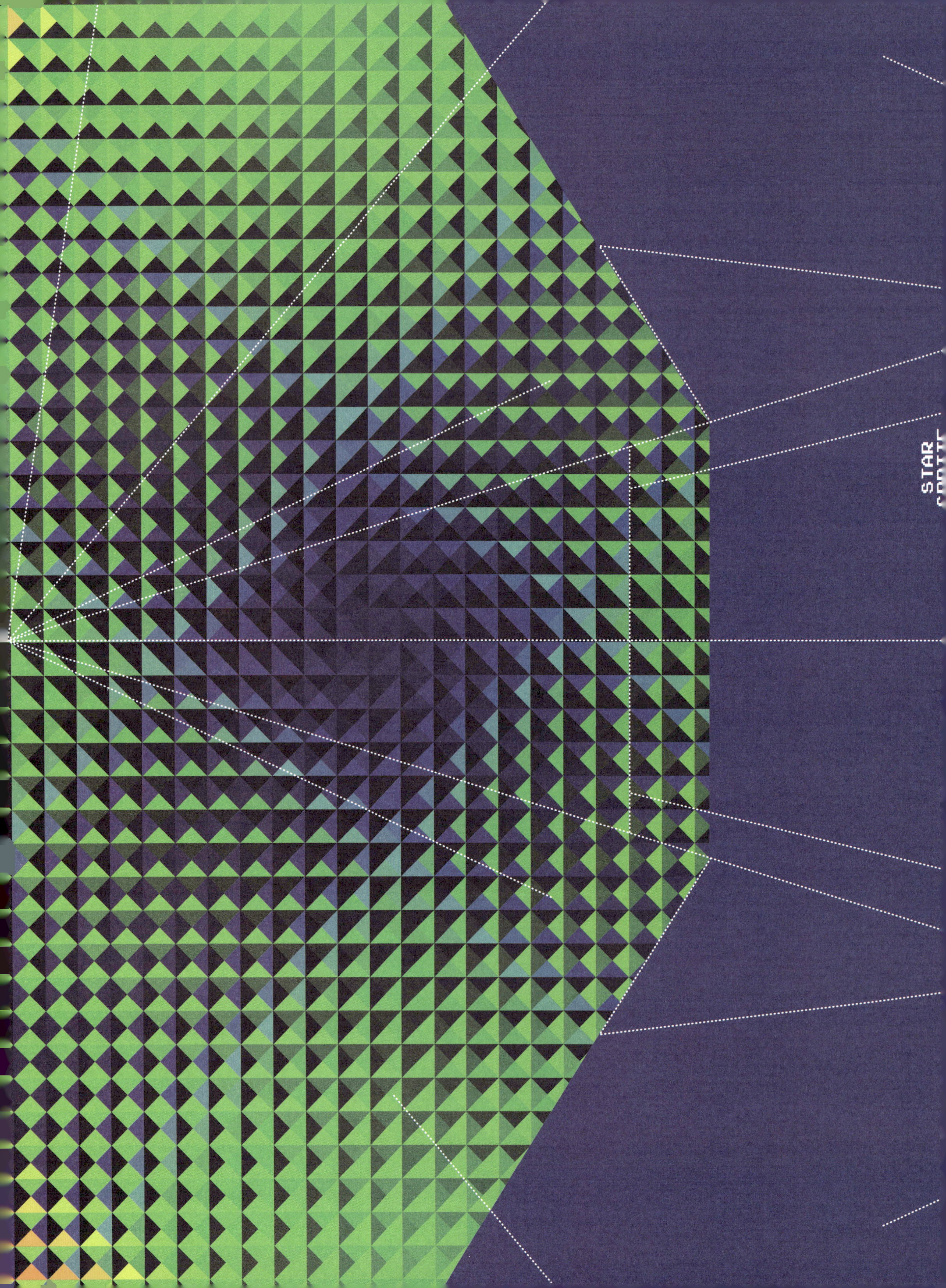
STAR

STAR

Mean Streak

Mean Streak
Mean Streak

Mean Streak
Mean Streak

Mean Streak

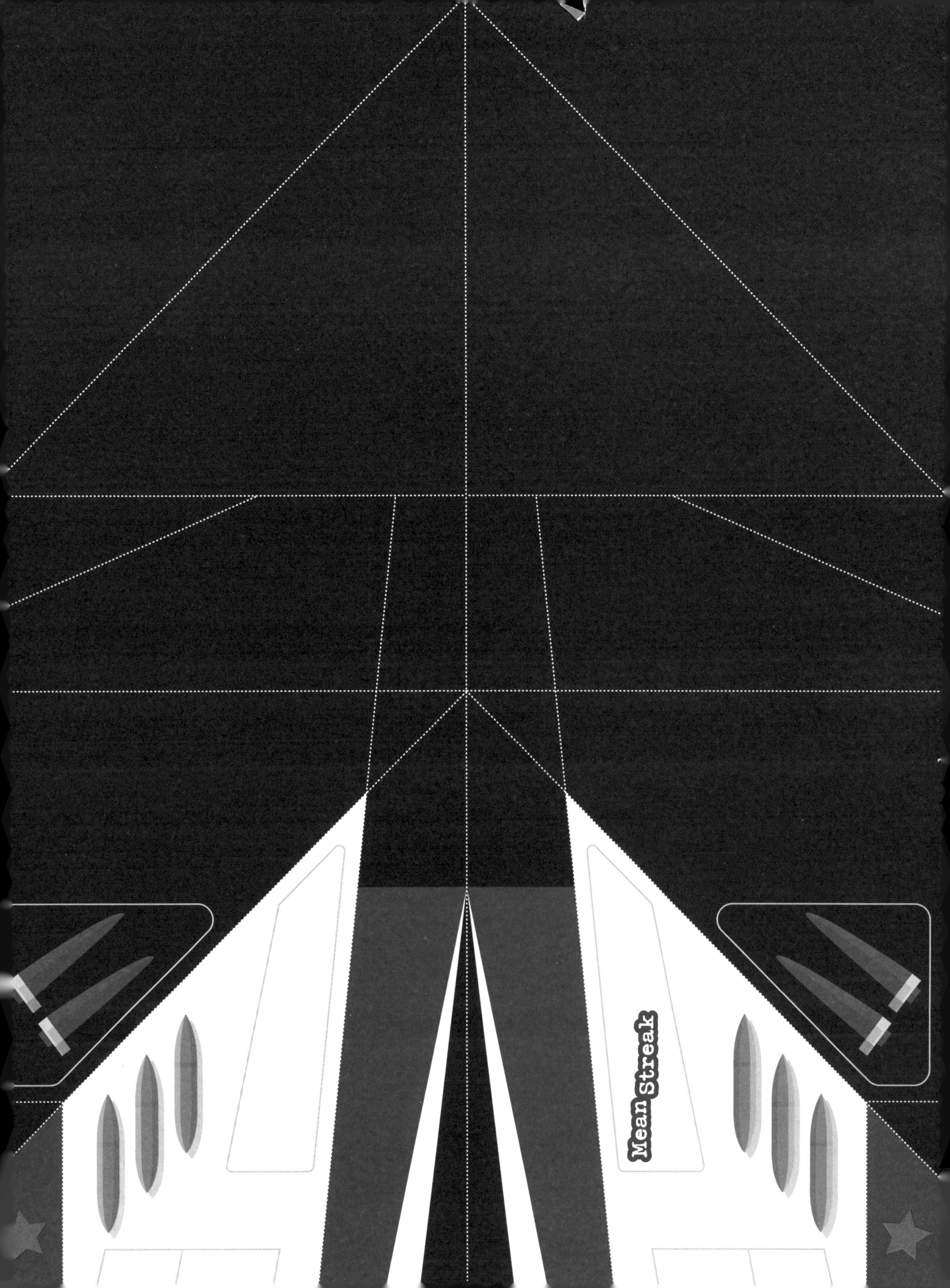
Mean Streak